Applied Physical Geography

GEOSYSTEMS IN THE LABORATORY
EIGHTH EDITION

CHARLES E. THOMSEN
AMERICAN RIVER COLLEGE

ROBERT W. CHRISTOPHERSON
AMERICAN RIVER COLLEGE – EMERITUS

Prentice Hall

Boston   Columbus   Indianapolis   New York   San Francisco   Upper Saddle River
Amsterdam   Cape Town   Dubai   London   Madrid   Milan   Munich   Paris   Montréal   Toronto
Delhi   Mexico City   São Paulo   Sydney   Hong Kong   Seoul   Singapore   Taipei   Tokyo

Geography Editor: Christian Botting
Marketing Manager: Maureen McLaughlin
Assistant Editor: Kristen Sanchez
Editorial Assistant: Bethany Sexton
Marketing Assistant: Nicola Houston
Managing Editor, Geosciences and Chemistry: Gina M. Cheselka
Project Manager, Science: Maureen Pancza
Operations Specialist: Maura Zaldivar
Supplement Cover Designer: Paul Gourhan
Cover Photo Credit: *Kvinberget Peak, south of Minebutka, Duvefjorden, Spitsbergen, Arctic Ocean. Dipping strata red-brown quartzites, weathered to produce talus slopes,* Bobbé Christopherson.

Printed in the United States of America

10 9 8 7 6 5 4 3 2 1

ISBN-13:   978-0-321-73214-9

ISBN-10:   0-321-73214-6

Prentice Hall
is an imprint of

www.pearsonhighered.com

Contents

Correlation Grid

Applied Physical Geography 8e	Geosystems 8e	Elemental Geosystems 7e	Elemental Geosystems 6e
Prologue Exercise	1, 8, Part 3	1, 5, Part 3	1, 5, Part 3
Exercise 1	1	1	1
Exercise 2	1	1	1
Exercise 3	1	1	1
Exercise 4	2	2	2
Exercise 5	5	3	3
Exercise 6	5	3	3
Exercise 7	3	2	2
Exercise 8	8	5	5
Exercise 9	8	5	5
Exercise 10	9	6	6
Exercise 11	10	7	7
Exercise 12	11, 12	8, 9	8, 9
Exercise 13	14	11	11
Exercise 14	1	1	1
Exercise 15	14	11	11
Exercise 16	17	13	14
Exercise 17	15, 16	12	12, 13
Exercise 18	13	10	10
Exercise 19	18	14	15
Exercise 20	19, 20	15, 16	16
Exercise 21	1	1	1

Preface

and

Personal Geography I.D.

Welcome to *Applied Physical Geography–Geosystems in the Laboratory*! During the eighth edition of this lab manual important scientific research and global trends will unfold. Physical geography, as the essential Earth systems spatial science, integrates the latest information and discoveries from across the globe. This is an exciting time to be taking a lab class in which many of the principles of physical geography are applied in specific exercises. In these exercises you have an opportunity to work with practices and tools of geographic analysis.

For example, imagine that toward the end of this century the climate of Illinois in the summer will be more like that of present-day east Texas—that's the climate change forecast! You get to work with this forecast and its implications in Lab Exercise 10.

Our goal in this laboratory manual is to help you use geographic principles, methods, and tools to better relate your physical geography text to understanding Earth's systems. A specific source for figures and materials used in this manual is *Geosystems—An Introduction to Physical Geography*, by Robert W. Christopherson, Pearson. Another source by this author is *Elemental Geosystems*, although this lab manual does not follow the specific chapter sequence of either text.

The Science of Geography

Geography (from *geo*, "earth," and *graphein*, "to write") is the science that studies the interdependence among geographic areas, natural systems, society, and cultural activities over space. As a discipline that synthesizes knowledge from many fields, geography integrates spatial elements to form a coherent picture of Earth. The term "spatial" refers to the nature and character of physical space: to measurements, relations, locations, and the distribution of things. Geography is governed by a method—a spatial approach—rather than by a specific body of knowledge.

Physical geography, therefore, centers on the spatial analysis of all the physical elements and processes that comprise the environment: energy, air, water, weather, climate, landforms, soils, animals, plants, and Earth itself. Physical geography, as a spatial human-Earth science, is in a unique position among sciences to synthesize and integrate the great physical and cultural diversity facing us. Aspects of physical geography requiring spatial analysis are in the news daily for these

are dramatic times relative to human-Earth relations and spatial change. Only through relevant education can an informed citizenry learn about the life-sustaining environment that surrounds and infuses our lives.

As a science, geography uses the scientific method. A focus study in Chapter 1 of both *Geosystems* and *Elemental Geosystems* briefly discusses this approach to problem solving. As you work through the laboratory manual, keep in mind this method of analyzing a problem, gathering data, organizing thought, and achieving the discovery of important principles.

Each Lab Exercise is organized in the following sequence to help you. Each exercise and its steps will take varying lengths of time depending on the emphasis taken in lab.

Organization of each Lab Exercise chapter:

- **Introduction**
  A brief overview of the subject and key terms and concepts involved in the exercise.
- **Key Terms and Concepts**
  Principal terms and concepts used in the exercise that are defined in the glossary and that you may want to check in a physical geography textbook.
- **Objectives**
  The skills and knowledge you should have on completion of the exercise.
- **Materials/Sources Needed**
  Items that you will need to complete the exercise.
- **Lab Exercise and Activities**
  ✳SECTION 1 headings such as this denote parts of the exercise and include specific numbered questions and activities to complete.

Your laboratory manual is designed for maximum utility—an asset to your learning activities. The standard symbols used on topographic maps appear inside the front cover. Inside the back cover are two world maps: the Köppen climate classification system and global biomes as Earth's terrestrial ecosystems. A fold-out flap, which you can deploy while you are working with the manual, features common metric-English conversions to assist you in this transition era between the two systems. The outside portion of the foldout flap presents the four classes of map projections. The back cover features a genetic climate classification map, which identifies causal elements that produce the pattern of world climates.

Note: The exercises in this manual are set up in a manner sensitive to the fact that many of the U.S. students still use English units, whereas Canadian students use metric measures. Conversions are given throughout the manual wherever appropriate. Your instructor will guide you as to which system of units to use in the exercises. The graphs and other items are designed to use either system although metric is preferred.

The initial laboratory exercise, called the Prologue Lab, is unique and does not appear in other lab manuals. Your lab instructor will explain the items for you to complete. Our experience in teaching geography is that students appreciate these tasks in the Prologue because they can experience the relevancy of geography. The Prologue Lab contains some activities that last throughout the term and not be completed until the last class, acting as threads through the lab. We have also included a section with Google Earth™ activities. If you have not used this wonderful tool, we hope you find it useful and fascinating. If you have used Google Earth™ before, we hope you enjoy this new use for it. Some of these tasks work best as group activities as guided by your instructor.

Preface to the Instructor

Applied Physical Geography: Geosystems in the Laboratory, is for the student taking an introductory physical geography course, either concurrently or previously. Included are background materials, brief text explanations, and figures from the *Geosystems* and *Elemental Geosystems* texts, although this laboratory manual is usable with other current textbooks in physical geography.

A separate and complete *Answer Key* for this laboratory manual is available to adopters of this lab manual. Be sure and ask your Pearson sales representative (name on file at your bookstore), or download it from www.pearsonhighered.com.

In the United States, contact the College Sales Department, 1 Lake Street, Upper Saddle River, NJ 07458, URL: **http://www.pearsonhighered.com.**

In Canada, contact Pearson Canada, 26 Prince Andrew Place, Don Mills, Ontario M3C 2T8, 1-800-263-9965 URL: **http://www.pearsoned.ca/;** or E-mail: **webinfo.pubcanada@pearsoned.com.**

Desk copies of *Geosystems, Elemental Geosystems*, or *Applied Physical Geography* will be sent to you immediately upon your request.

Course Equipment: We made a deliberate effort to write a manual requiring a minimum of equipment to keep your costs low. Students should provide their own pencils, colored pencils, metric-English ruler, protractor, calculator, and dictionary. Along with the

supplies and equipment you have on hand for labs to supplement this manual, the following items are needed: *Goode's World Atlas* (Rand McNally and Company), sling psychrometer, a compass, weather instruments, and soil text supplies, if available. You will also need stereo lenses in order to properly view the stereopair images in the manual. A sample stereopair photo is in the Prologue Lab for your students, with instruction to introduce this tool to them. There are nine sets of stereo photos in the lab manual and several stereo contour maps, demonstrating many concepts. Portions of 12 topographic sheets are included in the manual. Of course, you can supplement this brief presentation with your favorite topographic maps and stereoscopic and aerial photos featuring local examples. This edition features links to Google Earth™ KMZ files for several exercises so that students can actually fly through and experience 3-D landscapes as they work problems. The exercises can be completed in a campus computer lab if Google Earth™ is installed, or outside of class if facilities are not available on campus.

The manual begins with a unique Prologue Lab. This initial lab includes several activities that span the entire term. You may choose only portions of the activities presented. These elements were designed for continuity and to help you initiate *group activities* early in the class, especially the "Hazard Identification: Natural and Anthropogenic" assignment.

The twenty-one exercises are subdivided into ✷SECTION segments. Depending on your emphasis, a lab meeting may include several sections as part of an entire exercise, leaving out other sections as you see fit. We have designed the sections so that they can stand alone in most cases.

Building on the success of previous editions, this eighth edition follows the same format and sequence of exercises. The topographic maps have all been revised for improved clarity and vibrance. Improved writing with tighter, more focused student questions, and explanations are features of this edition. We continue the practice of placing some completed answers in each section to guide the student. We improved questions by requiring more interpretation of concepts under discussion. Please note there are new materials, work, and examples in the lab exercises on global climates, recurrence intervals for natural events, coastal and arid geomorphology, and a new geographic information systems exercise with data available for download, among many others. Note also the continuing improvement in production values with arts, maps, and overall reproduction.

The features that made the previous editions such a success are retained: the Prologue Lab as a term-length activity drawing the students together as a continuity thread; the stereopair photos, and stereo contour lines; the selection of color topographic maps and related exercises;

the "section" structure that allows you, the teacher, to edit and assign portions within a lab and the exclusion of other sections; and other features. We still remain the only lab manual with a complete glossary for student reference of concepts and with learning objectives with every lab. And for your convenience, we updated and improved the *Answer Key* with all the work completed, illustrations drawn, calculations made, and questions answered for quick reference when grading work.

Acknowledgments

We extend our continuing gratitude to the editorial, production, and sales staff of Pearson. Thanks to Christian Botting, Geography Editor, for willingness to risk new ideas and technologies, for his role in both *Geosystems* and *Elemental Geosystems*, and in geographic education in general. Our continuing compliments to the expertise of Assistant Editor Kristen Sanchez who oversees this *Applied Physical Geography* lab manual for her attention to detail and making the schedule work. And, our thanks go to the staff at Pearson Prentice Hall for publishing what you hold in your hands.

Robert acknowledges the continuing invaluable production and photographic work of his wife Bobbé. Charlie offers thanks for the wonderful support from his wife Holly and his children Emma and Finn in completing this work. Both of us acknowledge the role of students in our lives and the learning crucible that is the classroom.

In addition to all those geographers that contacted us by e-mail or at professional meetings, to Gail L. Hobbs and her work on several editions, and James Kernan for his careful and thoughtful review of the manuscript, we thank the following colleagues for the time and effort they invested in content reviews and work on previous editions (a combined alphabetical list).

Maura Abrahamson, Morton College
Mark R. Anderson, University of Nebraska, Lincoln
Gregorey D. Bierly, Indiana State University
Mark Blumler, SUNY, Binghamton
Woonsup Choi, University of Wisconsin-Milwaukee
Peter U. Clark, Oregon State University
John Dassinger, Chandler-Gilbert Community College
Lisa DeChano-Cook, Western Michigan University
Tracy Galarowicz, Central Michigan University
Robert H. Gorcik, William Rainey Harper College
Sharon Johnson, San Francisco State University
Cecil S. Keen, Mankato State University
James Kernan, SUNY Geneseo
Anil Kumar K Gangadharan, Georgia State University

Ingrid Luffman, East Tennessee State University
Joy Mast, Carthage College
Raoul Miller
Abdullah F. Rahman, Indiana University
Philip P. Reeder, University of Nebraska, Omaha
Bradley C. Rundquist, University of North Dakota
Donald C Rundquist, University of Nebraska
Susan Slowly, Blinn College
Tak Yung Susanna Tong, University of Cincinnati
Eugene Turne, California State University, Northridge
Forrest Wilkerson, Minnesota State University
Thomas Williams, Western Illinois University
Henry J. Zintambilla, Illinois State University
Matthew R. Zorn, Carthage College

Despite this expertise and support, we accept responsibility for whatever errors persist and for those areas needing further development toward future editions. We hope that *Applied Physical Geography* teaches principles effectively and enriches your experience with the science of physical geography. As it states in *Geosystems*, this is a remarkable moment in the human-Earth experience to be enrolled in physical geography.

Conclusion and Contact Information

We live in an extraordinary era of **Earth systems science**. This science contributes to our emerging view of Earth as a complete entity—an interacting set of physical, chemical, and biological systems that produce a whole Earth. Physical geography is at the heart of Earth systems science as we answer the spatial questions concerning Earth's physical systems and their interaction with living things. Hopefully, this lab is a step along the path of understanding Earth.

Contact information: As before, please consider the eighth edition of *Applied Physical Geography* an evolving work in progress. Given this admission, we welcome you to send any feedback, suggestions, criticisms, and comments to help us improve future editions. **Thank you for any assistance!**

Send comments and inquiries to:
Charles E. Thomsen,
American River College,
4700 College Oak Drive,
Sacramento, CA 95841
E-mail: thomsec@arc.losrios.edu

Robert W. Christopherson,
P. O. Box 128
Lincoln, CA 95648
E-mail: bobobbe@aol.com

A Personal Geography I.D.

The First Assignment

To begin: complete your personal Geography I.D. (see Preface page ix). Use the maps in a physical geography text, an atlas, college catalog, and additional library materials if needed. Your laboratory instructor will help you find additional source materials for data pertaining to the campus.

On text or atlas maps find the information requested noting the January and July monthly values (the small scale of such maps will permit only a general determination); air pressures are indicated by isobars, annual temperatures indicated by the isotherms, precipitation indicated by isohyets, climatic region, landform class, soil order, and ideal terrestrial biome, and the other information requested. Record the information from source maps in the spaces provided. The completed page gives you a relevant geographic profile of your immediate environment. As you progress through your physical geography class and physical geography lab, the full meaning of these descriptions will unfold. This page might be one you want to keep for future reference.

As you contemplate further schooling in colleges and universities in other parts of the world, or perhaps when you are preparing to move to another city, you may want to prepare a Geography I.D. for these new locations as part of your investigation into future options!

Geography I.D.

NAME: _____ LAB SECTION: _____

HOME TOWN: _____ LATITUDE: _____ LONGITUDE: _____

COLLEGE/UNIVERSITY: _____

CITY/TOWN: _____ COUNTY (PARISH): _____

Standard Time zone (for College Location): _____

Latitude: _____ Longitude: _____

Elevation (include location of measurement on campus): _____

Place (tangible and intangible aspects that make this place unique): _____

Region (aspects of unity shared with the area; cultural, historical, economic, environmental):

_____

_____

Population: Metropolitan Statistical Area (CMSA, PMSA, if applicable): _____

Environmental Data: (Information sources used: _____)

  January Avg. Temperature: _____ July Avg. Temperature: _____

  January Avg. Pressure (mb): _____ July Avg. Pressure: _____

  Average Annual Precipitation (cm/in.): _____

  Avg. Ann. Potential Evapotranspiration (if available; cm/in.): _____

Climate Region (Köppen symbol and name description): _____

_____

Main climatic influences (air mass source regions, air pressure, offshore ocean temperatures, etc.)

_____

Topographic Region or Structural Region (inc., rocks type, loess units, etc.): _____

_____

Dominant Regional Soil Order _____

Biome (terrestrial ecosystems description; ideal and present land use): _____

_____

_____

Prologue Lab

Physical Geography Logs, Hazard I.D. Group Activity, Weather Calendar, Google Earth™, and Stereophotos

Welcome to the Prologue Lab! All of us are geographers at one time or another. Our mobility is such that we cover great distances every day. Understanding both our lives and the planet's systems requires the tools of *spatial analysis* that are at the core of geography.

We personally interact with physical systems that are expressed in complex patterns across Earth. We need to develop continuity to our observations.

Throughout this term you may be using this Prologue Lab to record your observations of various environmental phenomena. Lab exercises that follow in later chapters explain various aspects of these observations in more detail. Your lab instructor will guide you as to which "Prologue" sections are required and the schedule for completion. These assignments are meant to span the entire term.

Key Terms and Concepts

environmental events log
hazard identification: natural and anthropogenic
hazard perception

weather and weather calendar
stereolenses
stereophotos

Objectives

After completion of this lab you should be able to:

1. *Analyze* Internet, printed, and published media coverage of environmental events to *discern* the spatial implications of these happenings.
2. *Relate* these contemporary events to your physical geography textbook, lecture discussions, and Lab, and to your daily experiences.
3. *Determine* natural and anthropogenic hazards on a local and regional scale.
4. *Identify* and *utilize* local sources of environmental data including broadcast and published media, instrument installations, and government sources to *complete* a weather calendar for up to 30 days of the term, as assigned by your instructor.
5. Use Google Earth™ to open placemarks and navigate through landscapes.
6. *Utilize* stereolenses to view photo stereopairs and use the Bright Angel, AZ, area for practice.

Environmental Events Log

These are challenging times of significant global change relative to the environment, societies, and cultures, and it is hoped that this exercise will bring these important subjects to light! A 2005 U.N. Environment Programme report that included 15 of the largest financial institutions, estimated that weather-related damage losses (drought, floods, hail, tornadoes, derechos, tropical systems, storm surge, blizzard and ice storms, and wildfires) will exceed $1 trillion by A.D. 2040 (adjusted to current dollars)—up from $210 billion in 2005 (Hurricane Katrina alone produced $125 billion in losses)–far exceeding the annual average of barely $2 billion, and the former record $90 billion of 1998. 2010 was the warmest year ever recorded, followed by 2005, 2007, 2009, 2002 and 1998 tied for the fifth warmest year, and every year of the last ten years is among the eleven warmest years ever.

Use the spaces provided to record environmental events: weather events (heavy rain, drought, winds, freezes), earthquakes, volcanic eruptions, floods and coastal inundation, tsunami events (seismic waves), biodiversity issues and species extinctions, land-slides, record icebergs, Antarctic ice-shelf disintegration, stratospheric ozone updates, air pollution occurrences, or other significant events involving the physical elements of the environment. Place the number of the item in its correct location on the accompanying map(s) at the end of this log; perhaps color coding the number according to the type of event (blue for weather, brown for geomorphology, green for biogeography, etc.). A classroom bulletin board map might be used to compile a class world/North American map of environmental events.

Please access available media (Internet, broadcast, newspapers and magazines, and journals) as time permits. Use public television (PBS) and National Public Radio (NPR) shows such as NOVA, Lehrer NewsHour, National Geographic, Nature, and National Public Radio's "Morning Edition" and "All Things Considered," or any other sources. *Geosystems*, seventh edition, lists more than 200 URLs for Internet sources. The Internet is particularly rich for learning about the latest occurrences.

You will see how much of physical geography is actually occurring out there in "the real world." Your list is not meant to be comprehensive, rather it should reflect a growing interest in the physical systems that surround our lives. Restrict your list to events that occur throughout the term or semester. Space for eighteen items are provided.

---

1. Date   **Source**   **Event**

_____ ; _____ ; _____

_____

_____

Tie-in to physical geography: _____

2. Date   **Source**   **Event**

_____ ; _____ ; _____

_____

_____

Tie-in to physical geography: _____

---

3. Date   Source   Event

_____ ;   _____ ;   _____

_____

_____

Tie-in to physical geography: _____

4. Date   Source   Event

_____ ;   _____ ;   _____

_____

_____

Tie-in to physical geography: _____

5. Date   Source   Event

_____ ;   _____ ;   _____

_____

_____

Tie-in to physical geography: _____

6. Date   Source   Event

_____ ;   _____ ;   _____

_____

_____

Tie-in to physical geography: _____

7. Date   Source   Event

_____ ;   _____ ;   _____

_____

_____

Tie-in to physical geography: _____

8. Date   Source   Event

_____ ;   _____ ;   _____

_____

_____

Tie-in to physical geography: _____

9. <u>Date</u>   **<u>Source</u>**   **<u>Event</u>**

_____; _____; _____

_____

_____

Tie-in to physical geography: _____

10. <u>Date</u>   **<u>Source</u>**   **<u>Event</u>**

_____; _____; _____

_____

_____

Tie-in to physical geography: _____

11. <u>Date</u>   **<u>Source</u>**   **<u>Event</u>**

_____; _____; _____

_____

_____

Tie-in to physical geography: _____

12. <u>Date</u>   **<u>Source</u>**   **<u>Event</u>**

_____; _____; _____

_____

_____

Tie-in to physical geography: _____

13. <u>Date</u>   **<u>Source</u>**   **<u>Event</u>**

_____; _____; _____

_____

_____

Tie-in to physical geography: _____

14. <u>Date</u>   **<u>Source</u>**   **<u>Event</u>**

_____; _____; _____

_____

_____

Tie-in to physical geography: _____

15. Date   **Source**   **Event**

_____; _____; _____

_____

_____

Tie-in to physical geography: _____

16. Date   **Source**   **Event**

_____; _____; _____

_____

_____

Tie-in to physical geography: _____

17. Date   **Source**   **Event**

_____; _____; _____

_____

_____

Tie-in to physical geography: _____

18. Date   **Source**   **Event**

_____; _____; _____

_____

_____

Tie-in to physical geography: _____

Extra—Date   **Source**   **Event**

_____; _____; _____

_____

_____

Tie-in to physical geography: _____

Figure P.1
Location of environmental events in the world (locate and label):

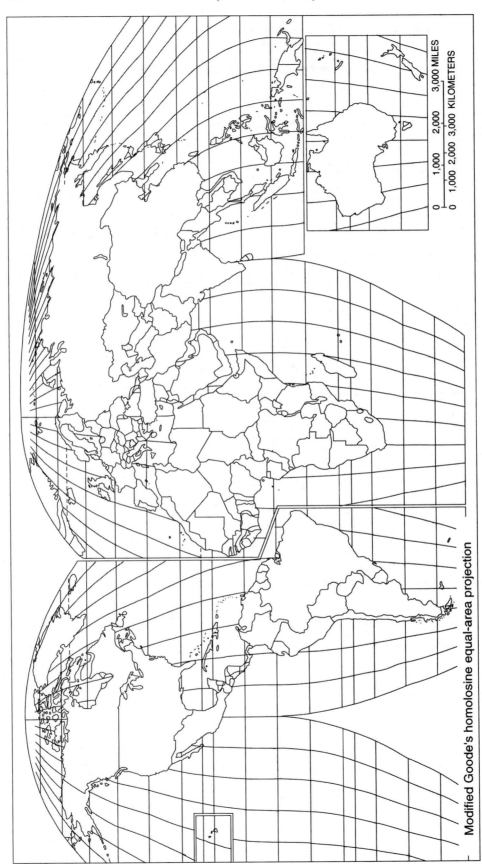

Modified Goode's homolosine equal-area projection

North America
(Political boundaries and rivers)

0   500   1000 Kilometers

0   300   600   Miles

Figure P.2
Location of environmental events in the United States and Canada (locate and label):

Hazard Identification:

Natural hazards and Anthropogenic (human-forced) hazards
—A Group Activity—

A valid and applicable generalization is that humans are unable or unwilling or incapable of perceiving hazards in a familiar environment. Such an axiom of human behavior certainly helps explain why large populations continue to live and work in hazard-prone settings. Similar questions also can be raised about populations in areas vulnerable to natural disasters. Political and economic interests ride this wave of poor public perception and lack of physical geography's spatial perspective. Ideally, an informed public that acts on its knowledge can move to prevent development in hazardous areas. This poor perception applies to human-induced hazards as well, such as groundwater contamination, air pollution, and radiological hazards.

A consequence of poor hazard perception is that our entire society bears the financial cost, the subsequent bad planning, and irrational development—no matter where in the country it occurs. Of course, those directly affected are the victims who must shoulder the physical, emotional, and economic hardship of the disaster, and for which society indirectly pays. This recurrent, yet avoidable cycle—construction, devastation, reconstruction, devastation—continues. Recent events are reminders of this problem in our relationship with nature. A few examples include: the European heat wave of 2003 that killed over 40,000, the Russian heat wave that killed 15,000 and; the devas-

tating tornadoes of the last decade, with more than 1600 hitting the U.S. alone in 2004 the Midwest floods of 1993, 1995, and 1997; the nor'easters that struck the East Coast during January 1998 and January 2005; Hurricanes Katrina, Rita, and Wilma in 2005; Cyclone Sidr in 2007; the 2001 Gujarat, India, earthquake that killed 20,000, or the 30,000 that died in Bam, Iran, from a 2003 earthquake, the October 2005 earthquake in Pakistan that killed 80,000, the China earthquake of 2008, and the Haiti earthquake of 2010 that killed over 230,000; or the more than 225,000 killed by the tsunamis caused by the magnitude 9.0 earthquake near the northwest coast of Sumatra, Indonesia; the 2010 eruption of Eyjafjallajökull that disrupted air travel for months; and the devastating 2011 9.0 earthquake off of Japan, followed by the tsunami, and the nuclear power plant meltdowns; among many. The increasing pace of global climate change demands that we pay attention to our relations with Earth systems.

A value in studying physical geography is a better knowledge of the environment and the workings of natural systems. This heightened awareness may lead to a more reasonable assessment of natural and anthropogenic hazards. Following group discussions with lab partners, describe and explain any natural or human-induced hazards that exist within these radii from campus. Note how they affect you.

10 km (6 mi) radius of your campus:

Natural– _____

_____

_____

_____

_____

Anthropogenic– _____

_____

_____

_____

100 km (60 mi) radius of your campus:

Natural– _____

_____

_____

_____

_____

Anthropogenic– _____

_____

_____

_____

_____

Your state or province:

Natural– _____

_____

_____

_____

_____

Anthropogenic– _____

_____

_____

_____

_____

Additional specific regional environmental issues:

_____

_____

_____

_____

Name: _____   Laboratory Section:_____

Date: _____   Score/Grade: _____

Weather Calendar

Weather is an important integrative subject within physical geography. Temperature, air pressure, relative humidity, wind speed and direction, day length, and Sun angle are important measurable elements that contribute to the weather. **Weather** is the short-term condition of the atmosphere, as compared to **climate**, which reflects long-term atmospheric conditions and extremes.

We go online to the Internet or tune to a local station for the day's weather report from the National Weather Service (in the United States, **http://www.nws.noaa.gov**) or the Canadian Meteorological Center (in Canada, **http://www.msc-smc.ec.gc.ca/contents_e.html**) to see the current satellite images and to hear tomorrow's forecast. Most cable systems carry the *Weather Channel*, which presents local weather information throughout the day (**http://www.weather.com**). Internationally the World Meteorological Organization coordinates weather information (see **http://www.wmo.ch/**). Many sources of weather information and related topics are found on the *Geosystems* Home Page, under "Destination" in Chapter 8 (Chapter 5 in *Elemental Geosystems*).

Although this is early in the term and weather has not specifically been discussed in lecture, you will find this exercise an interesting, ongoing work-in-progress that will stimulate many questions and prepare you for those chapters when they are covered.

Meteorology is the scientific study of the atmosphere (*meteor* means "heavenly" or "of the atmosphere"). Embodied within this science is a study of the atmosphere's physical characteristics and motions; related chemical, physical, and geological processes; the complex linkages of atmospheric systems; and weather forecasting. Computers permit the handling of enormous amounts of data on water vapor, clouds, precipitation, and radiation for accurate forecasting of near-term weather.

New developments in supercomputers and ground- and space-based observation systems are making this a time of dramatic developments. A few of the innovations include: wind profilers that use radar to profile winds from the surface to high altitudes, computer-based modeling software for 3-D weather models and displays, improved international cooperation and weather data dissemination, and a standard Advanced Weather Interactive Processing System (AWIPS) at more than 150 stations. Data gathering is enhanced by the completed installation of 883 Automated Surface Observing System (ASOS) instrument arrays as a primary surface weather-observing network. By the end of 2007, 158 WSR-88D (Weather Surveillance Radar) Doppler radar systems as part of the NEXRAD (next Generation Weather Radar) program were operational through the National Weather Service, in conjunction with the Federal Aviation Administration and the Department of Defense. In Canada, the National Radar Project will have 30 CWSR-98 radars in service.

NOAA operates the Forecast Systems Laboratory, Boulder, Colorado, which is developing new forecasting tools (see **http://www.fsl.noaa.gov/**). Research and monitoring of violent weather is centered at NOAA's National Severe Storms Laboratory in several cities (see **http://www.nssl.noaa.gov**). For tropical weather see the National Hurricane Center at **http://www.nhc.noaa.gov/**.

Building a database of atmospheric conditions is key to *numerical weather prediction* and the development of weather forecasting models. *Synoptic analysis* involves the characterization of weather elements over a region within the same time frame, as shown on the weather map presented in LAB EXERCISE 9 of this manual. You will not be observing all these elements for your weather calendar, but a complete weather forecast is based on knowing the following 9 items (the 7 marked with an asterisk are required for this exercise).

- Barometric pressure*
- Pressure tendency
- Surface air temperature*
- Dew-point temperature*
- Wind speed and direction*
- Type of clouds*
- Current weather*
- State of the sky*
- Precipitation since last observation

The purpose of this weather exercise is twofold. *First*, to help you find sources of weather information. Our lives take place within an environment of variable weather, so it is important to have access to such information.

Second, for you to gather data for up to 30 days (or less, as assigned by your instructor) during the term, not necessarily consecutive days, and record them on a weather calendar using official international symbols and designations. If your lab meets 15 or 16 times during the term and the department has weather instruments, then half your days of observation are easily completed on class day. The other days can be met by outside sources or visits to the department.

The National Weather Service and the Canadian Meteorological Center use a standardized system of recording and reporting weather data around a weather station symbol developed by the World Meteorological Organization (WMO). An example of this method is shown below; a complete explanation of symbols and data forms is in LAB EXERCISE 9.

WEATHER STATION SYMBOL

WIND SPEED

WIND DIRECTION

TEMPERATURE °F

PRESSURE

74   1004

PRECIPITATION → ✲✲
               0

DEW POINT °F

CLOUD COVER

COLD FRONT   WARM FRONT

STATIONARY FRONT   OCCLUDED FRONT

PRECIPITATION TYPE

,	DRIZZLE
•	RAIN
✱	SNOW
▽	SHOWERS
⊺⌐	THUNDERSTORMS
≡	FOG
∞	DRY HAZE
⌒⌣	FREEZING RAIN
△	HAIL
⚠	SLEET

WIND SPEED
(kph/mph)

CALM	⊚	
5–13		3–8
14–23		9–14
24–32		15–20
33–40		21–25
89–97		55–60
116–124		72–77

CLOUD COVER

◯	NO CLOUDS	◑	SIX-TENTHS
◔	ONE-TENTH OR LESS	◕	SEVEN-TENTHS TO EIGHT-TENTHS
◔	TWO-TENTHS TO THREE-TENTHS	◗	NINE-TENTHS OR OVERCAST WITH OPENINGS
◔	FOUR-TENTHS	●	COMPLETELY OVERCAST
◑	FIVE-TENTHS	⊗	SKY OBSCURED

Attached is your "Weather Calendar" with spaces for up to 30 days of weather observations. These days need not be consecutive and may be spread out over the entire term. The completed calendar is due at the end of the term. Your instructor will give you more details. The circle in the middle of each square is the weather station symbol, above and below are spaces to list the date and the place of observation (if you are using a broadcast source, list the place the weather instruments are located that made the observation).

Weather station symbol explanations are in Figure 9.1 in LAB EXERCISE 9. Let's refer to the weather station symbol as if it was a clock. Temperature is recorded in the 10:00 position, barometric pressure at 2:00, weather type at 9:00, dew-point temperature (if available) at 8:00, wind speed and direction varies in location depending on the wind itself. If you are observing clouds: high clouds are symbolized above the station, middle and low clouds below the station symbol.

The state of the sky (cloud cover) is the area inside the circle. In the example: temperature is 74°F (23°C), pressure is 1004.0 mb, the sky is 50% (five-tenths) overcast, with slight continuous fall of snowflakes at time of observation. Winds are from the northwest (line drawn on the side of the station where the winds originate) at 15–20 mph (24–32 kmph) as noted by the "flags" on the line. For a complete listing of weather symbols see Lab Exercise 9.

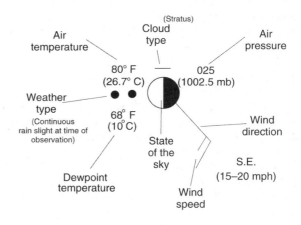

Weather Calendar

Date/time_____ ○ Place_____	Date/time_____ ○ Place_____	Date/time_____ ○ Place_____	Date/time_____ ○ Place_____	Date/time_____ ○ Place_____
Date/time_____ ○ Place_____	Date/time_____ ○ Place_____	Date/time_____ ○ Place_____	Date/time_____ ○ Place_____	Date/time_____ ○ Place_____
Date/time_____ ○ Place_____	Date/time_____ ○ Place_____	Date/time_____ ○ Place_____	Date/time_____ ○ Place_____	Date/time_____ ○ Place_____
Date/time_____ ○ Place_____	Date/time_____ ○ Place_____	Date/time_____ ○ Place_____	Date/time_____ ○ Place_____	Date/time_____ ○ Place_____
Date/time_____ ○ Place_____	Date/time_____ ○ Place_____	Date/time_____ ○ Place_____	Date/time_____ ○ Place_____	Date/time_____ ○ Place_____
Date/time_____ ○ Place_____	Date/time_____ ○ Place_____	Date/time_____ ○ Place_____	Date/time_____ ○ Place_____	Date/time_____ ○ Place_____

Google Earth™

Virtual geography in *Applied Physical Geography*

Incorporated in this edition of the Lab Manual are exercises built around Google Earth™. This technology is the most exciting and useful geographic tool since GPS (Global Positioning System) technology became available to the public. This program allows you to view a photo-realistic 3-D model of Earth. You can zoom in and out, tilt and rotate the view, adjust the vertical exaggeration of the landscape, and add your own image overlays. Some of the exercises in the Lab Manual will involve simply navigating around terrain features to gain a better understanding of how they look, while other exercises will involve the application of critical thinking skills.

To use this wonderful free tool you will need to download Google Earth™ from **http://earth.google. com**. The Google Earth™ web site has information on the hardware and network requirements for the program. It also has an excellent tour of the features of the program and a gallery of downloadable Placemarks called KMZ or KML files.

Once you have installed the program you should go through their tour of the program. The tour covers the basics of how to navigate around the virtual world, how to search for places, and how to open and save Placemarks.

The exercises in this manual will all involve downloading KMZ files from the Geosystems website and opening them in Google Earth™. This introduc-tory exercise will help acquaint you with the process of opening a KMZ file and using it with the questions of a lab exercise. The first KMZ file is called in-tro.kmz. For the KMZ file and the exercise questions go to **http://mygeoscienceplace.com**. This web page will have the KMZ files and the questions for each assignment. Depending on how your web browser is set up, the KMZ file will automatically open up in Google Earth™ or you may have to download it to your local machine and open it up manually.

Once you have opened the KMZ it will appear in the sidebar in the Places box, in the temporary places folder. When you double-click on the Intro Placemark it will fly you to the first location and open up a pop-up window with information about that place. Some of the Placemarks will only have one location; others may have several locations that you will visit, as well as overlay images that you can turn on and off. You can control the transparency of an image overlay with the transparency slider control. This will be useful when you wish to interpret information from multiple overlays. After you have found your house (which is the first thing almost everyone does), go through each of the questions in the Intro assignment. The questions for each assignment are found on the same website as the KMZ files. Your instructor will tell you how to submit your assignments.

Aerial Photo Stereopairs

3-D photography in this manual

If you cover one eye and look about you, you see length and width but not depth. The view with one eye is *monocular vision*. The normal depth perception we experience is because we use two eyes and see with *binocular vision*, or *stereoscopic vision*. Each eye sees objects at a slightly different angle. Our brain blends these two signal inputs into a three-dimensional image that features the dimensions of length, width, and depth.

The photo stereopairs in this manual are made by a camera platform that takes aerial photos from slightly different positions along a flight path. The United States Geological Survey's National Aerial Photography Program (NAPP) is the source for the photographs we use, all at a 1:40,000 scale.

Along north-south NAPP flight paths, adjoining frames of aerial photographs overlap by 60%. In these overlap areas the camera views the same scene from different angles. When we place any two overlapping photos side-by-side and view them through stereolenses, or a stereoscope, each eye assumes the position of the camera as it took each picture. These stereolenses straighten your line of sight (to infinity) so that each eye looks ahead to one of the photographs. If you relax your eyes, the overlapping area of the photographs will spring forth in three dimensions. This stereoscopic technique assists in landscape analysis and enhances photo interpretation.

To give you some practice with this technique we present a demonstration photo stereopair for you. On the next page is a pair of NAPP photographs showing the Bright Angel area, in the Grand Canyon of Arizona. We chose this scene because the topographic relief is dramatic and the Grand Canyon is a familiar place. Because of the north-south flight path, these photos are oriented in such a way that north is to the right, south to the left, and west at the top of the page.

Hold the lenses to your eyes and relax. If you are doing it correctly, you will see 2 light-colored lines (the single margin between the photos). As your eyes adjust, the area that appears to be between the lines is where you will see stereovision. Experiment until you find the right distance from the photos, usually about 25 cm, 10 in., above the photo pair. Initially, stereoscopic viewing might take a few minutes to adjust your eyes; give yourself time.

The vertical depth that you see is several times more than is actually in Bright Angel Canyon. This is called vertical exaggeration and is one of the traits of stereophotography. Vertical exaggeration helps when viewing some scenes of gentle, rolling topography and low relief.

In the stereophoto observe, the intricately dissected Colorado Plateau cliffs, slopes, vista points, and a deep inner V-shaped canyon occupied by Bright Angel Creek. Study the landscape and compare your results with others in your lab section.

For more information about NAPP, go to the USGS and the EROS Data Center at **http://edc.usgs.gov/ products/aerial/napp.html**.

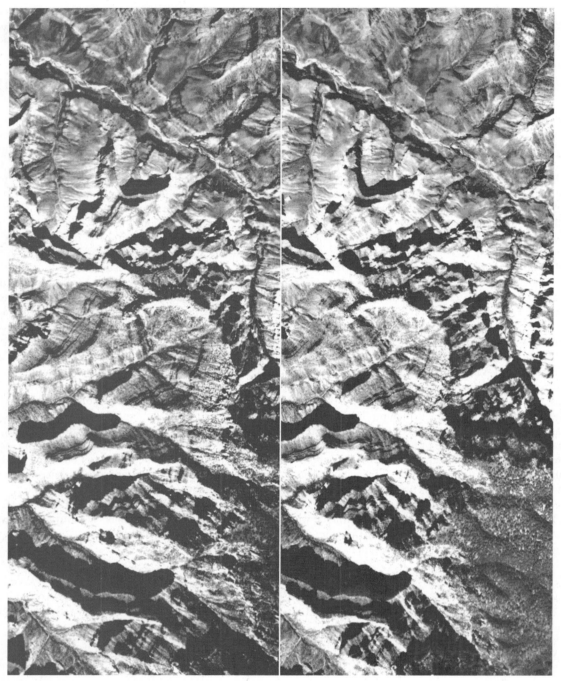

Figure P.3
Photo stereopair of a portion of Bright Angel Canyon, Arizona. Note that north is to the right, south to the left, and west to the top. (NAPP photos by USGS, EROS Data Center.)

Name: _____   **Laboratory Section:** _____

Date: _____   **Score/Grade:** _____

Lab Exercise 1

Latitude, Longitude, and Time

To know specifically where something is located on Earth's surface, a coordinated grid system is needed, one that is agreed to by all peoples. The terms *latitude* and *longitude* were in use on maps as early as the first century A.D., with the concepts themselves dating back to Eratosthenes and others. As you complete your *Geography I.D.* in the Preface, you will eventually determine the time zone you live in and the absolute location in latitude and longitude for your campus. A knowledge of longitude and Earth's rotation forms the basis of standard time and the world system of Coordinated Universal Time (UTC). We examine both this essential geographic location grid and time in this exercise. Lab Exercise 1 features six sections.

Key Terms and Concepts

antipode
Coordinated Universal Time (UTC)
daylight saving time
geographic grid
great circle
Greenwich Mean Time (GMT)
International Date Line

latitude
local Sun time
longitude
meridian
parallel
prime meridian
small circle

Objectives

After completion of this lab you should be able to:

1. *Define* a great circle and a small circle and *describe* the relationship between a great circle and travel.
2. *Define* latitude and parallel, longitude and meridian, and *use* them in simple sketches to *demonstrate* how Earth's reference grid is established.
3. *Use* latitude and longitude to *locate* places on Earth's surface.
4. *Relate* the time where you are with world standard time and *calculate* differences in standard and Sun time.
5. *Contrast* the prime meridian with the International Date Line.

Materials/Sources Needed

globe
string
ruler
protractor
world atlas

Lab Exercise and Activities

Great and Small Circles

Great circles and small circles are important concepts relating to latitude and longitude; concepts that we will refer to in later activities. A **great circle** is any circle of Earth's circumference whose center coincides with the center of Earth. An infinite number of great circles can be drawn on Earth. On flat maps, airline and shipping routes appear to arch their way across oceans and landmasses following great circles. Despite their curved appearance on flat maps, great circle routes are the shortest distance between two points on Earth.

Using a globe and a piece of string, place one end on San Francisco and stretch the string tautly to London. This direct route is a portion of a great circle. **Small circles** are circles whose centers do not coincide with Earth's center. In Figure 1.1 properly draw and label a great circle and a small circle, other than those noted by the latitude and longitude grid.

Figure 1.1
Label great circles and small circles

Again using the globe and piece of string, show the great circle routes linking the following cities, listing at least four or five prominent geographical features (e.g., mountains, seas, rivers, nations, cities, etc.) that you would cross as you traveled along those routes:

1. London, England and Colombo, Sri Lanka (English Channel, Germany, Carpathian Mtns.) _____

   _____

2. Vancouver, British Columbia and Sydney, Australia _____

   _____

3. Your home town and Beijing, China _____

_____

✳ SECTION 2

Latitude and Parallels

Adapting from the Babylonians, Ptolemy divided the circle into 360 *degrees* (360°), with each degree subdivided into 60 *minutes* (60′), and each minute further subdivided into 60 *seconds* (60″). This method of dividing degrees into minutes and seconds is often referred to as DMS coordinates. Geographic Information Systems, GPS, and other computer-based systems often use decimal degrees or DD coordinates. In this method, each degree is divided into a more familiar base ten system. We'll now examine each of these grid coordinate elements.

   Latitude is an angular or **arc distance** north or south of the *equator* (the line running east to west halfway between the poles), measured from the center of Earth (Figure 1.2a). On a map or globe, the lines designating these angles of latitude run east and west, parallel to the equator, whose latitude is 00.00°.

   The North Pole is 90.00° north latitude, and the South Pole is 90.00° south latitude. "Lower latitudes" are those nearer the equator, whereas "higher latitudes" refer to those nearer the poles.

   A line connecting all points at the same latitudinal angle is called a **parallel**. Only one parallel—the equatorial parallel—is a great circle; all others are small circles, diminishing in circumference north and south toward the poles, which are merely points. As shown in Figures 1.2a and 1.2b, *latitude is the name of the angle* (49° north latitude), *parallel names the line* (49th parallel), and both indicate arc distance north of the equator.

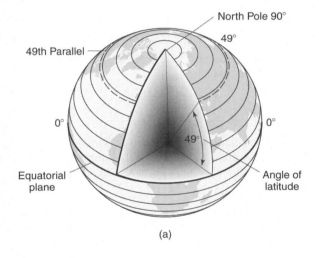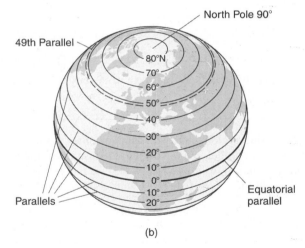

(a)   (b)

Figure 1.2 a and b
Latitude and parallels

1. Figure 1.3 on the following page is a half-Earth section through the poles. An angle of 30° has been measured north of the equator, with the 30th parallel drawn on the globe. Following this example, use your protractor to measure an angle of 23.5° north of the equator and draw and label the corresponding parallel, which is known as the Tropic of Cancer. Measure another angle 66.5° north of the equator, draw and label the parallel, known as the Arctic Circle. Repeat this in the Southern Hemisphere, drawing and labeling the Tropic of Capricorn (23.5° south) and the Antarctic Circle (66.5° south). We will be working with these parallels while studying Earth-Sun relationships in Lab Exercise 4. Lastly, measure, draw, and label one last parallel to show your own latitudinal location.

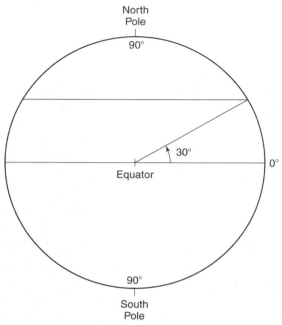

North
Pole
90°

30°

0°

Equator

90°

South
Pole

Figure 1.3
Measuring latitudes on Earth

2. Parallels have often been used to demarcate political boundaries. The 49th parallel north forms a portion of the border between which two countries?

_____

3. Parallels made famous by wars in the last century include the _____ parallel dividing the two

Koreas and the _____ parallel that divided Vietnam until 1975. (The border was approxi-

mately 80 km (50 mi) north of the city of Hue.)

_____

4. Using a globe and your atlas or other world map, locate three cities that are located at approximately the 23rd parallel in the Northern Hemisphere; note their location in degrees (and minutes if your map is detailed enough to estimate minutes). Use the globe first, then refer to the atlas maps to better determine specific latitudes. Be sure and list their country names as well.

<u>**City and Country Name**</u>   <u>**Longitude (degrees and minutes if possible)**</u>

a) _____   _____

b) _____   _____

c) _____   _____

5. Now locate three cities that are located at approximately your latitude.

<u>**City and Country Name**</u>   <u>**Longitude**</u>

a) _____   _____

b) _____   _____

c) _____   _____

Longitude and Meridians

Longitude is an angular or **arc distance** east or west of a point on Earth's surface, measured from the center of Earth (Figure 1.4a). On a map or globe, the lines designating these angles of longitude run north and south at right angles (90°) to the equator and to all parallels. A line connecting all points at the same lon-gitude is a **meridian**. Every meridian is one-half of a great circle that passes through the poles. *Longitude is the name of the angle, meridian names the line*, and both indicate distance east or west of an arbitrary **prime meridian**.

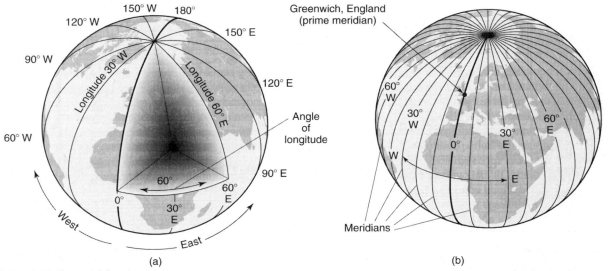

Figure 1.4 a and b
Longitude and meridians

Figure 1.5 is a view of Earth from directly above the North Pole; the equator is the full circumference around the edge. (Using a globe, look at it from above the North Pole to see this perspective.) A line has been drawn from the North Pole to the equator and labeled 0°, representing the prime meridian. Earth's prime meridian through Greenwich, England, was not generally agreed to by most nations until 1884. To the right of 0° on the diagram is the *Eastern Hemisphere*—label this—and to the left of 0° is the *Western Hemisphere*—label this.

Extend another line from the North Pole to the other side of Earth, opposite the prime meridian, and label it 180°. You now have marked the line that is the International Date Line, which extends from North to South Poles on the opposite side of Earth from the prime meridian. Using your protractor, measure, draw, and label the meridi-ans that are 100° east and 60° west of the Greenwich meridian. Lastly, locate, draw, and label the meridian that marks your present longitude.

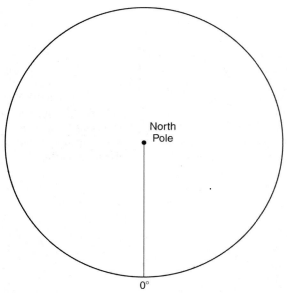

North
Pole

0°

Figure 1.5
Measuring longitudes on Earth

1. On a political globe or world map follow the International Date Line across the Pacific Ocean. Why do you

   think the International Date Line is not straight, but zigs and zags? _____

   _____

2. Examine an atlas or a political globe and in the spaces marked a through h list the provinces and states
   through which the 100th meridian in the Western Hemisphere passes—north to south. The first answer is
   provided for you in bracketed italics.

   _____

   a) _____*[Nunavut, Canada]*_____   e) _____

   b) _____   f) _____

   c) _____   g) _____

   d) _____   h) _____

The 98th meridian is roughly the location of the 51 cm (20 in) isohyet (a line connecting points of equal precipitation), with wetter conditions to the east and drier conditions to the west. In North America, tall grass prairies once rose to heights of 2 m (6.5 ft) and extended westward across the Great Plains to about the 98th meridian, with short-grass prairies farther west. The deep sod formed beneath these tall grasslands posed problems for the first settlers, as did the climate. The self-scouring steel plow, introduced in 1837 by John Deere, allowed the interlaced grass sod to be broken apart, freeing the soils for agriculture. Keep this information about the region around the 100th meridian, and close to the 98th in mind as a "location reference" later on when studying precipitation and vegetation.

Earth's Geographic Grid

Latitude and longitude form Earth's geographic grid. Every location on our planet is described with this Cartesian coordinate system—where location is described on a plane by two intersecting lines. The GEOGRAPHY I.D. asks you to identify the general coordinates of your home town and college campus. *Antipodes* are points on the globe that are diametrically opposite each other that would be at either end of a line drawn through the center of Earth—such as the North Pole and South Pole. Determine the following geographic grid coordinates and cities from an atlas, wall map, or large political world globe. While wall maps and globes will help you see points in relation to the rest of the world, you should use regional maps in an atlas to be more precise in your readings.

1. Locate and give the geographic coordinates for the following cities (to a tenth of a degree if your atlas maps are detailed enough) or identify the cities from the given coordinates.

City	Latitude and Longitude
a) Greenwich, London, England	[51.5°N 0°]
b) Rio de Janeiro, Brazil	
c) Your state's/province's capital city	
d) _____ [Tokyo, Japan] _____	35.7°N 139.7°E
e) _____	8.8°S 13.2°E

On the map grid in Figure 1.6, plot the coordinates in items 1 (a) through (e) above; and label the city names.

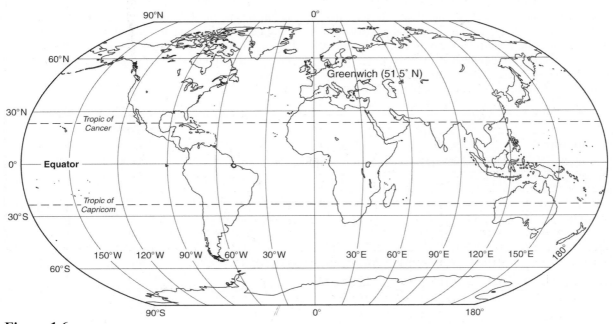

Figure 1.6
Plotting coordinates

2. Using your knowledge of latitude and longitude, find and circle the errors in the following geographic grid coordinates. Rewrite the coordinates correctly in the space to the right. You do not have to locate these on a map.

    a) DMS format Lat. 57° 86′ 24″ S, Long. 149°02′63″ N   _____   [*minutes cannot exceed 60*]

    b) DD format Lat. 105.03° W, Long. 93.99° E   _____

3. If you were halfway between the equator and the South Pole and one-quarter of the way around Earth to the west of the Prime Meridian, what would be your latitude and longitude?

    _____

4. You are at 10° N and 30° E; you move to a new location which is 25° south and 40° west of your present location. What is your new latitudinal/longitudinal position?

    _____

5. You are at 20° S and 165° E; you move to a new location which is 45° north and 50° east from your present location. What is your new latitudinal/longitudinal position?

    _____

✳ SECTION 5

Latitude and Longitude Values

Because meridians of longitude converge toward the poles, the actual distance on the ground (**linear distance**) spanned by a degree of longitude is greatest at the equator (where meridians separate to their widest distance apart) and diminishes to zero at the poles (where they converge). The linear distance covered by a degree of latitude, however, varies only slightly, owing to the oblateness of Earth's shape. Table 1.1 compares the linear or physical distance of a degree of latitude and longitude at selected latitudinal locations. It also shows the similarity of ground distance for a degree of latitude and a degree of longitude at the equator.

	Latitude Degree Length		Longitude Degree Length			
Latitudinal Location	**km**	**(mi)**	**km**	**(mi)**		
90° (poles)	111.70	(69.41)	0	(0)	North Pole	
60°	111.42	(69.23)	55.80	(34.67)	1° of latitude = 111.70 km	1° of longitude = 0 km
50°	111.23	(69.12)	71.70	(44.55)	1° of latitude = 111.04 km	1° of longitude = 85.40 km
40°	111.04	(69.00)	85.40	(53.07)	40°	
30°	110.86	(68.89)	96.49	(59.96)	1° of latitude = 110.58 km (1 arc of a meridian)	Equator
0° (equator)	110.58	(68.71)	111.32	(69.17)	1° of longitude = 111.32 km (1 arc of a parallel)	

Table 1.1
Physical Distances Represented by Degrees of Latitude and Longitude

1. From the table, you can see that latitude lines are evenly spaced, approximately 111 km (69 miles) apart at any latitude. Using these values as the linear distance separating each degree of latitude, the distance between any given pair of parallels can be calculated. (*Note: Locations must be due north-south of each other.*) For example, Denver is approximately 40° north of the equator (arc distance). The linear distance between Denver and the equator can be calculated as follows:

   40° N to 0° = 40° × 111 km/1° = 4440 km

   or

   40° N to 0° = 40° × 69 miles/1° = 2760 miles

   Using these same values for a degree of latitude and an atlas for city location, calculate the linear distance in km and miles between the following sets of points (along a meridian):

   a) Mumbai, India and the equator _____

   b) Miami, Florida and 10° south latitude _____

   c) Edinburgh, Scotland and the 5th parallel north _____

   d) Your location and the equator _____

2. The table also shows that the linear distance separating each 1° of longitude decreases toward the poles. For example, at 30° latitude each degree of longitude is separated by slightly more than 96 km (nearly 60 miles), and at 60° latitude, the linear distance is reduced to approximately half that at the equator. For each of the following latitudes, determine the linear distance in km and in miles for 15° of longitudinal arc (along a parallel):

   km   **miles**

   a) 40° latitude: _____ *[15°×85.40 km = 1281 km]* _____   _____ *[15°×53.07 = 796 mi]* _____

   b) 50° latitude: _____   _____

   c) 90° latitude: _____   _____

3. What do you estimate is the approximate linear distance of a degree of longitude at your present location in km? _____ in miles? _____ one minute of longitude in km? _____ in miles? _____ one second of longitude in km? _____ in miles? _____ a tenth of a degree in km? _____ in miles? _____ one hundredth of a degree in km? _____ in miles? _____ (longitudinal arc along a parallel).

4. Again using Table 1.1, what is the linear distance in km and miles along the parallel at your latitude from the prime meridian to you? _____

Time, Time Zones, and the International Date Line

Today we take for granted international standard time zones and an agreed-upon prime meridian. If you live in Oklahoma City and it is 3:00 P.M., you know that it is 4:00 P.M. in Baltimore and Toronto, 2:00 P.M. in Salt Lake City, and 1:00 P.M. in Seattle and Los Angeles. You also probably realize that it is 9:00 P.M. in London and midnight in Riyadh, Saudi Arabia. (The designation A.M. is for *ante meridiem* for "before midday/noon," whereas P.M. is for *post meridiem*, meaning "after midday/noon.") Coordination of international trade, airline schedules, business activities, and daily living depends on a common time system.

Because Earth revolves 360° every 24 hours, or 15° per hour $(360° \div 24 = 15°)$, a time zone of one hour is established for each 15° of longitude. Each time zone theoretically covers 7.5° on either side of a controlling (standard) meridian (0°, 15°, 30°, 45°, 60°, 75°, 90°, 105°, 120°, etc.) and represents one hour.

Traveling eastward, one would set the clock one hour later, and going westward, one hour earlier for each time zone.

Local Sun time, or solar time, is based on the actual longitudinal arc distance (in degree, minutes, and seconds) between a location and the prime meridian. There could be as much as 30 minutes difference in time (or more, depending on the actual time zone boundaries) between local Sun time and standard time.

Opposite the globe from the prime meridian is the **International Date Line**, the starting point for each new day. Since the Date Line is the standard (controlling) meridian for a time zone, no change in time occurs when crossing the Line, only the date. When traveling westward, one will set the calendar one day ahead, and when crossing the Line heading eastward, the calendar will be reset one day back.

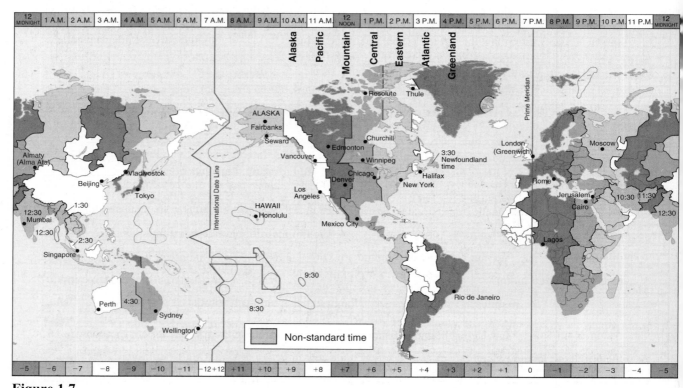

Figure 1.7

Modern international standard time zones. The numbers at the bottom indicate how many hours each zone is earlier (plus sign) or later (minus sign) than Greenwich Mean Time, today known as Coordinated Universal Time (UTC) or Zulu Time.

1. From the map of global time zones in Figure 1.7, determine the present time in the following cities: (For your time use the starting time of the lab.)

Moscow _____   Los Angeles _____

London _____   Honolulu _____

Chicago _____   Mumbai _____

2. You may not always have a time zone map available, but by remembering the relationship of 1 hour for every 15° of longitude, you can easily calculate the difference in time between places. <u>Indicating and using the standard meridians to determine time zones</u>, solve the following problems. Show your work:

   a) If it is 3 A.M. Wednesday in Vladivostok, Russia (132° E), what day and time is it in Moscow (37° E)?
   [The controlling meridian for Moscow is 105° away from Vladivostok's controlling meridian of 135° E
   (135° − 30° = 105° difference). Since Earth rotates 15° per hour, Moscow is 7 hours earlier than
   Vladivostok (105° difference / 15° rotation per hour = 7 hours time difference), therefore if it is 3 A.M.
   Wednesday in Vladivostock it is 8 P.M. Tuesday in Moscow.]

   b) If it is 7:30 P.M. Thursday in Winnipeg, Manitoba, Canada (97° W), what day and time is it in Harare, Zimbabwe (31° E)? _____

   c) If you depart from San Francisco International Airport at 10:00 P.M. on Tuesday, what day and time will you arrive in Auckland, New Zealand (175° E), assuming a flight time of 14 hours?

   _____

3. If there is a difference of 15° of longitude for each hour of time, how much difference in time is there for 1° of longitude? _____

   for 1′ of longitude? _____

4. What is the standard (controlling) meridian for your time zone (75°—Eastern, 90°—Central, 105°—Mountain, 120°—Pacific, 135°—Alaska, other)?

   _____

   How many degrees of longitude separate you from this standard controlling meridian?

   _____

   How does your distance from the standard meridian affect the difference between the time on your clocks and actual Sun (solar) time? (Calculate the difference between standard and Sun time using the answer you determined in #3 above). _____

   _____

   Knowing your standard meridian is useful because it will tell you how many minutes off (fast or slow) your watch is in your time zone from the actual position of the Sun in the sky.

Just as you can use longitudinal distance to calculate time, you can also use time difference to calculate your longitude. This is the method that was often used to determine longitudinal location when sailing (if they didn't happen to have a high technology GPS—Global Positioning System—on board!). Again, the relationship between time and longitude (15° = 1 hour of time) is the key. If you know the difference in time between two places, and you know the longitude of one place, you can calculate the longitude of the other. For example: Your watch (kept on your hometown standard time) reads 10:30 A.M., and you see that a clock in the town you are visiting reads 7:30 P.M. Your home near St. Paul, Minnesota, uses 90° W as its standard meridian. Therefore:

- Time difference between 10:30 A.M. and 7:30 P.M. = 9 hours
- 9 hours × 15°/hour = 135° of longitude separating the two locations
- Since you are at a place later in time than home, you must be east of home
- 135° east of your hometown at 90° W would be 45° E longitude

5. Assume the time on your watch, showing local standard time, is 4:15 P.M. A chronometer (a clock giving Coordinated Universal Time) reads 2:15 A.M. What is your longitude? _____

_____

Daylight saving time is the practice of setting time ahead one hour in the spring and back one hour in the fall in the Northern Hemisphere. Clocks in the United States were left advanced an hour continuously from 1942 to 1945 during World War II, and again from January 1974 to October 1975, in order to conserve energy during an oil embargo. Daylight saving time increased in length in the United States and Canada in 1986. Under the Energy Policy Act of 2005, Daylight saving time begins three weeks earlier than previously, at 2 A.M. on the second Sunday in March. DST has been extended by one week and now lasts until 2 A.M. on the first Sunday of November.

6. Does your community adopt daylight saving time? What are the dates for adjusting clocks in the spring and fall? _____

7. What time does your physical geography lab start

   a) according to standard time? _____

   b) according to daylight saving time? _____

   c) in UTC? _____
        (24-hour clock time in Greenwich, England; e.g., 3:00 P.M. = 15:00 hours)

Lab Exercise 2

Directions and Compass Readings

Earth's coordinate grid system was covered in Lab Exercise 1, so you are familiar with using latitude and longitude to find a point's absolute location on Earth's surface. There are times when relative direction between two places is desired, for example if you want to navigate (over land, by air, or by water) from one point to another. There are three systems used to indicate relative direction on maps: 1) compass points, 2) azimuths, and 3) quadrant compass bearings.

Understanding and being able to use these direction-finding methods is important whether you are reading a street/road map or are interpreting specialized maps such as topographic maps which are discussed and used in later exercises. These systems are all measurements of horizontal direction, as if you were in the air looking down at Earth from above.

Compasses are useful for finding directions, but corrections must be made for deviation from true or geographic north. Lab Exercise 2 has five sections.

Key Terms and Concepts

azimuths
cardinal points
compass points
geographic north
grid north
isogons
isogonal map

magnetic compass
magnetic declination
magnetic north
quadrant compass bearings
topographic map
true north

Objectives

After completion of this lab, you should be able to:

1. *Describe* the methods of indicating direction using various types of compass readings and *apply* those methods to compass directions.
2. *Define* magnetic declination and *determine* differences in "north" readings on a map.
3. *Explain* any difference between a local topographic map and the isogonal map relative to magnetic declination.

Materials/Sources Needed

color pencils
compass
local topographic map
protractor
ruler

Lab Exercise and Activities

✳ SECTION 1

Compass Points

The directional system familiar to most people is that of **compass points**. The symbol that you find at the beginning of each new Lab Exercise in this manual is a *compass rose*. Cartographers often included elaborate compass roses as a decorative, as well as a functional part of antique maps. The compass rose has four *cardinal points*—in a clockwise direction—north, east, south, and west, separated by four intermediate points: NE, SE, SW, NW. These are split into 16 and then, sometimes, again into 32 compass points.

1. Label 16 compass points on Figure 2.1, using color pencils on the compass rose itself to distinguish the categories of division (include a legend for your color scheme). In each division the points are listed clockwise.

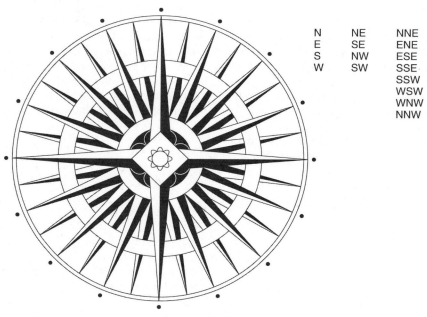

N	NE	NNE
E	SE	ENE
S	NW	ESE
W	SW	SSE
		SSW
		WSW
		WNW
		NNW

Figure 2.1
Compass points

Azimuths

While useful, compass points are not extremely accurate, and most often the direction between two points does not fall exactly on one of the 32 points. Degrees of an arc are used for more precise directions.

Azimuths are read from the north in a clockwise direction, from 0° to 360° (with 0° and 360° being the same point—north). For further precision the degrees can be broken down into minutes (').

1. Figure 2.2 has a few azimuths drawn and labeled as examples. Using your protractor, determine the azimuth readings for *A* and *B*, labeling the value for each on the diagram.

2. Measure, draw, and label the following azimuths on the diagram: 230°, 78°, 145°.

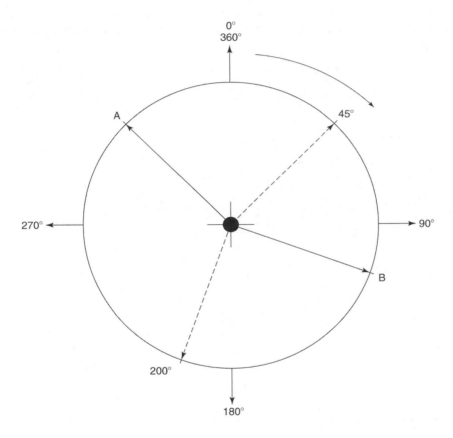

Figure 2.2
Azimuths

Quadrant Compass Bearings

Quadrant compass bearings, often referred to simply as *bearings*, divide the circle into four sections, or *quadrants*, so that the maximum bearing is 90°. The readings are measured eastward or westward from either a north or a south reference (whichever is closer to the bearing), and thus in either a clockwise or counterclockwise direction from 0° N or 0° S.

North and south are both 0°. In reading the bearings, both letters and numbers are used. In reading N 35° E, think "*(from)* north 35° *(toward)* east," and for S E, remember that the reading is "*(from)* south *(toward)* east." (When actually reading the bearings, don't use "*from*" and "*toward*.") You work further with azimuths and compass bearings in Lab Exercise 14 (topographic map reading).

1. Figure 2.3 has two bearings drawn and labeled as examples for you. The north bearing is 35° east from north; the south bearing is 85° east from south.

   a) Using your protractor, determine the bearings for **A** and **B**, labeling the values on the diagram.

   b) Measure, draw, and label the following bearings on the diagram: N 67° E, N 60° W, S 55° W, S 10° E.

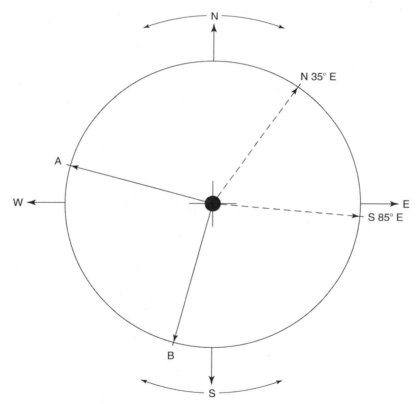

Figure 2.3
Quadrant compass bearings

2. Convert the azimuths values listed below to quadrant bearings:

127° [*S 53° E*] 24° _____ 305° _____ 197° _____ 85° _____

Convert the quadrant bearings listed below to azimuths values:

S 71° W [*251°*] N 10° W _____ N 43° E _____ S 30° E _____ S 10° W _____

✳ SECTION 4

Compass Declination

Earth rotates around an axis that is marked by the **geographic North** and **South Poles**, also referred to as **true North** and **true South**. Meridians of longitude converge at these geographical points.

A *magnetic compass*, an instrument used to determine direction, points to the **magnetic North Pole**, which does not coincide with true north and the geographic pole. The difference between the north arrow on a magnetic compass and true north is the **magnetic declination**. A compass needle that points west of true north indicates a *west declination*, and a compass nee-

dle that points east of true north indicates an *east declination*. When using a magnetic compass, you must make adjustments for the difference between magnetic north and true north.

Isogons, lines connecting points of equal magnetic declination, are shown on *isogonal maps* such as in Figure 2.4. The magnetic poles change over time, slowly migrating, so that adjustments for these magnetic pole changes need to be considered if map navigation is to be up-to-date.

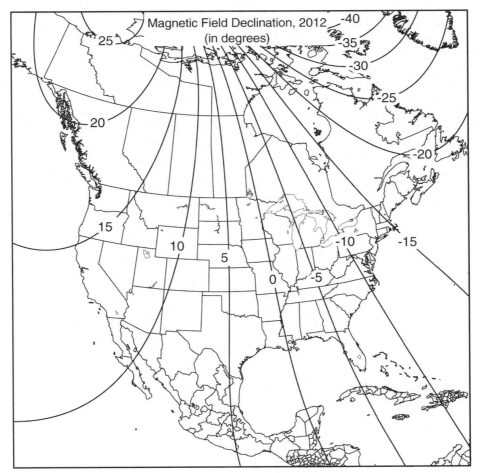

Figure 2.4
Isogonal map for 2012

1. Using the isogonal map, (Figure 2.4) and your atlas, in what city would you have to stand during 2012 to get a compass reading with zero magnetic declination (where magnetic north aligns with true north, or an agonic line). Mark the city with a dot and a label on the map.

2. Using your atlas, turn to the pages featuring the polar regions and find the latitude/longitude coordinates for the:

   a) Magnetic North Pole _____

   b) Magnetic South Pole _____

3. Find the linear distance in km and miles between the latitude of the:

   a) Magnetic North Pole and true North _____

   b) Magnetic South Pole and true South _____

Topographic maps (or topo maps) are a set of maps covering the entire United States that show the horizontal position (latitude/longitude) of boundaries, land-use aspects, bodies of water, and economic and cultural features. Topographic maps also have a vertical component to show topography (configuration of the land surface), including slope and relief (the vertical difference in local landscape elevation). These fine details are shown through the use of elevation contour lines. A contour line connects all points at the same elevation. Elevations are shown above or below a vertical datum, or reference level, which usually is mean sea level. The contour interval is the vertical distance in elevation between two adjacent contour lines. Inside the front cover of this manual are the standard symbols commonly used on these topographic maps. These symbols and the colors used are standard on all USGS topographic maps: black for human constructions, blue for water features, brown for relief features and contours, pink for urbanized areas, and green for woodlands, orchards, brush, and the like. The margins of a topographic map contain a wealth of information about its concept and content. In the margins of topographic maps, you find the quadrangle name, names of adjoining quads, quad series and type, position in the latitude-longitude and other coordinate systems, title, legend, datum plane, symbols used for roads and trails, the dates and history of the survey of that particular quad, magnetic declination (alignment of magnetic north) and compass information, and more.

Magnetic declination is shown by a declination diagram or arrow in the margin of a topographic map as shown in Figure 2.5. True north is indicated by the star, and the magnetic declination is shown by the arrow marked MN. You work further with magnetic declination in Lab Exercise 14 (topographic map reading). Grid north (GN) refers to the relation of the topo map to the Universal Transverse Mercator system (UTM). GN is another reference system based on rectangular map zones, so that as meridians taper toward the pole the GN shows the relation of these converging lines and the rectangular grid.

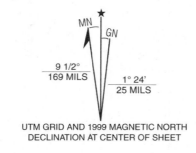

MN
GN

9 1/2°
169 MILS

1° 24'
25 MILS

UTM GRID AND 1999 MAGNETIC NORTH
DECLINATION AT CENTER OF SHEET

CUMBERLAND, MD

MARYLAND

QUADRANGLE LOCATION

Figure 2.5
Declination arrow on topographic maps

4. According to Figure 2.5, what is the magnetic declination between

    a) True north and magnetic north (MN)? _____

    b) True north and grid north (GN)? _____

5. Look for the declination arrow in the margin of the topographic map provided by your instructor. What is the magnetic declination between true north and magnetic north on the topographic map provided by your instructor? What is the magnetic declination between true north and magnetic north shown for this location in Figure 2.4? Any discrepancies you find relate to the migrating magnetic pole and the dates of the topo map and the isogonal map (2012). (Keep this topo map handy for the next sections.)

_____

_____

✳ SECTION 5

Compass Bearing

The compass bearing between two points on a map may be determined by using a protractor or a magnetic compass. To accomplish this, lightly draw a line through the specified points, extending the line to the edge of the map. To use a protractor, place the straight edge of the protractor along the margin of the map, with the origin of the protractor at the point where the line intersects the map margin. The extension of the drawn line indicates the compass bearing (Figure 2.6a). To use the compass, simply align the margin of the map with the 0°/360° and 180° points on the compass. The bearing can be read from the compass dial. (Pay no attention to the compass needle.) (Figure 2.6b).

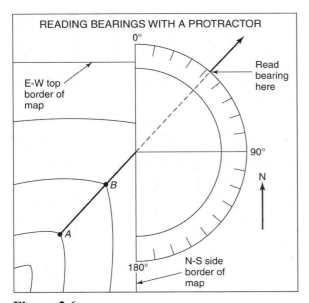
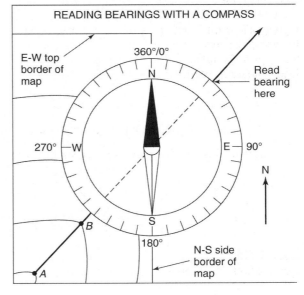

Figure 2.6
Reading compass bearings on a map using a protractor (a); using a magnetic compass (b). (From Busch, Richard M. editor, *Laboratory Manual in Physical Geology*, 3rd ed., Macmillan Publishing Company © 1993.)

Using the local topographic map from the last section, complete the following.

1. Find the compass bearing between two points on the map indicated by your instructor. Give the reading as an

   azimuth. _____ and

   as a quadrant compass bearing _____

   On the circles provided below, sketch your determination of azimuth and quadrant compass bearing.

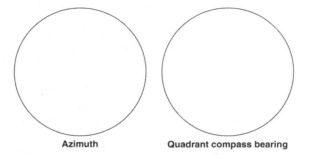

   Azimuth   Quadrant compass bearing

2. What is the highest elevation on the local topographic map? (Indicate a geographical feature that is at or

   adjacent to that elevation.) _____

   _____

3. What is the lowest elevation on the map? (Indicate a geographical feature that is at or adjacent to that elevation.)

   _____

   _____

4. What is the relief (highest elevation minus lowest elevation in a local landscape) of the area portrayed on

   this map? Perhaps compare two or three areas of the topographic map if relief varies across the region.

   (Show your work.) _____

   _____

Name: _____   Laboratory Section: _____

Date: _____   Score/Grade: _____

Lab Exercise 3

Map Projections, Map Reading, and Interpretation

A **map** is a generalized view of an area, usually some portion of Earth's surface, as seen from above and greatly reduced in size. The part of geography that embodies mapmaking is called **cartography**. Maps are critical tools with which geographers depict spatial information and analyze spatial relationships. We all use maps at some time to visualize our location and our relationship to other places, or maybe to plan a trip, or to coordinate commercial and economic activities.

In this lab exercise, we begin with an overview of various map projections that are used for a variety of purposes. An important decision in making maps is the proper scale to use. Maps, like architectural blueprints, are produced at a greatly reduced size compared to reality. Then, we specifically work with a U.S. mapping program—the township and range system. Lab Exercise 3 has four sections.

Key Terms and Concepts

cartography
conic
cylindrical
equal area (equivalent)
gnomonic projection
great circle
map

map projection
Mercator projection
planar
rhumb lines
scale
township and range
true shape (conformal)

Objectives

After completion of this lab, you should be able to:

1. *Define* the map projection concept and *identify* the four classes of map projections.
2. *Recognize* distortions that are characteristic of selected map projections and *compare* distortions in order to *select* the appropriate map for a given use.
3. *Plot* a great circle route from a gnomonic projection to a Mercator projection.
4. *Compare* methods of expressing map scale and *calculate* map scales.
5. *Explain* the township and range system employed in the U.S.

Materials/Sources Needed

tracing (or wax) paper or notebook paper
world globe
scissors
color pencils
calculator
ruler

Lab Exercise and Activities

✳ SECTION 1

Map Projections

We worked with Earth's coordinate grid system in Lab Exercise 1. Cartographers prepare large-scale flat maps—two-dimensional representations (scale models) of our three-dimensional Earth. This transformation of spherical Earth and its latitude-longitude coordinate grid system to a flat surface in some orderly and systematic realignment is called **a map projection.**

A globe is the only true representation of distance, direction, area, shape, and proximity, and preparation of a flat version requires decisions as to the type and amount of distortion that will be acceptable. To understand this problem, consider these important physical properties of a globe:

- parallels always are parallel to each other, always are evenly spaced along meridians, and always decrease in length toward the poles.
- meridians converge at both poles and are evenly spaced along any individual parallel.
- the distance between meridians decreases toward poles, with the spacing between meridians at the 60th parallel equal to one-half the equatorial spacing.
- parallels and meridians always cross each other at right angles.

The problem with preserving these properties is that all globe qualities cannot be reproduced on a flat surface. The larger the area of Earth depicted on the map, the greater the distortion will be. Simply taking a globe apart and laying it flat on a table illustrates the problem faced by cartographers in constructing a flat map (Figure 3.1).

In Figure 3.1, you can see the empty spaces that open up between the sections, or gores, of the globe when it is flattened. Flat maps always possess some degree of distortion—much less for large-scale maps representing a few kilometers, much more for small-scale maps covering individual countries, continents, or the entire world.

The best projection is always determined by its intended use. The major decisions in selecting a map projection involve the properties of **equal area** (equivalence) and **true shape** (conformality). If a cartographer selects equal area as the desired trait, as for a map showing the distribution of world climates, then shape must be sacrificed by *stretching* and *shearing* (have parallels and meridians cross at other than right angles).

If, on the other hand, true shape is desired, as for a map used for navigational purposes, then equal area must be sacrificed, and the scale will actually change from one region of the map to another. Two additional properties related to map projections are true direction and true distance.

The fold-out flap of the back cover shows four classes of map projections and geometric-surface perspectives from which three classes of maps—**cylindrical**, **planar** (or *azimuthal*), and **conic**—are generated. Another class of projections that cannot be derived from this physical-perspective approach is the non-perspective *oval*-shaped. Still others are derived from purely mathematical calculations.

With all projections, the contact line (or point) between the globe and the projection surface—called a **standard line**—is *the only place where all globe properties are preserved.* A *standard parallel* or *standard meridian* is true to scale along its entire length without any distortion. Areas away from this critical tangent line or tangent point become increasingly distorted. Consequently, this area of optimum spatial properties should be centered on the region of immediate interest so that greatest accuracy is preserved there.

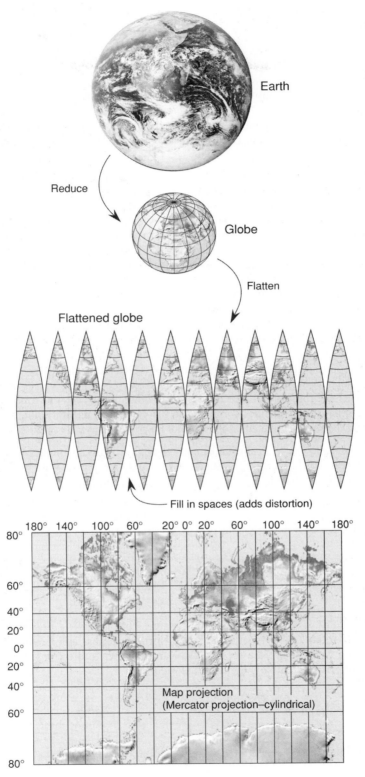

Figure 3.1
Conversion of a globe to a flat map requires decisions about the desired properties to preserve, how much distortion is acceptable, and which projection best serves the purpose of the map.

Cylindrical, Planar, and Conic Projections

This step may be done by working in groups. Your instructor will direct you.

1. Using a globe and tracing (or wax) paper—notebook paper works with an illuminated globe—trace outlines of North America, South America, and Greenland. You can do this quickly; a rough outline showing size and shape is sufficient.

   Using a globe and your tracings compare the relative sizes of North America, South America, and Greenland. Which is larger and by approximately how many times? Describe your observations.

   _____

   _____

   _____

2. Examples of two cylindrical projections are presented in Figure 3.2, the **Mercator projection** (conformal, true-shape); and Figure 3.3, the *Lambert projection* (equivalent, equal-area). These two map projections— the Mercator and the Lambert—are simply being used to examine *relative* size and shape in each portrayal of Earth.

   Trace around the outlines of the same continents and island as indicated above—North America, South America, and Greenland. You might want to mark and color code the tracings according to their source (Mercator or Lambert) for future reference. Use your tracings to answer the following questions. Keep in mind that these two sets of map outlines compared to the globe you used are not at the same scale: the distance measured along their respective equators will not be equal.

   a) Using the outlines from the Mercator and Lambert map projections on the next pages, make the same comparisons among North America, South America, and Greenland. Which is relatively larger and by approximately how many times? Again, describe your observations.

   Mercator _____

   _____

   Lambert _____

   _____

   b) Now compare the relative shapes of Greenland, North America, and South America from the globe (shows true shape) with those from the Mercator and Lambert projections. Is there distortion in terms of shape? If so, briefly explain how they are distorted.

   Mercator _____

   _____

   Lambert _____

   _____

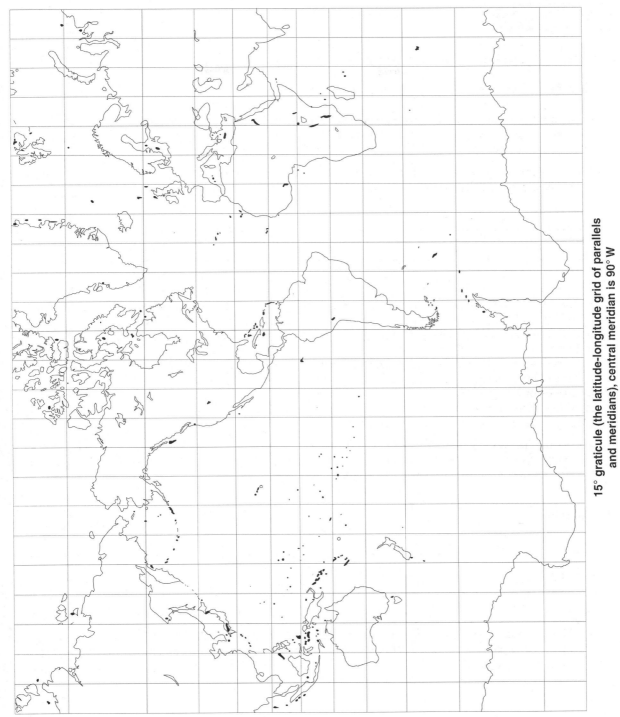

15° graticule (the latitude-longitude grid of parallels and meridians), central meridian is 90° W

Figure 3.2
Mercator projection—conformal, true-shape map

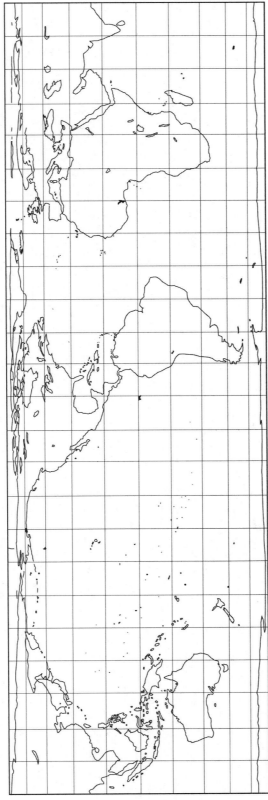

Figure 3.3
Lambert projection—equivalent, equal-area map

Figure 3.4a and Figure 3.4b once again show the Mercator and Lambert cylindrical projections. Diagrams to the right of each projection help you to identify the distortion in each. The circles along the equator—which is the standard line, or line of tangency between the globe and the cylinder and are without distortion—are the "reference standard." The shape and size of the other "circles" is proportionate to the distortion at various latitudes on the maps.

3. What happens to the circles in the Mercator projection (and, therefore, to the landmasses) as latitude increases? What causes this distortion of the circles away from the equator? (Hint: compare parallels and

meridians with those on the globe.) _____

_____

4. What happens to the circles in the Lambert projection (and, therefore, to the landmasses) as latitude increases? What causes this distortion of the circles away from the equator? (Once again, compare parallels

and meridians.) _____

_____

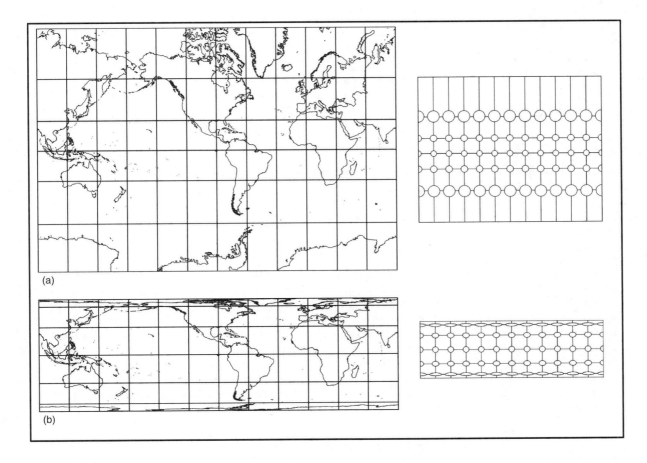

Figure 3.4
(a) Mercator cylindrical true shape (conformal) projection, 30° graticule the latitude-longitude grid; standard parallel 0° central meridian 90° W. (b) Lambert cylindrical equal area (equivalent) projection, 30° graticule; standard parallel 0° central meridian 90° W.

5. Cylindrical projections such as Mercator and Lambert would best be used in mapping what areas of the globe? Why? _____

_____

_____

6. Figure 3.5a shows Earth on a *planar* projection, in which the globe is projected onto a plane. This is a **gnomonic projection** and is generated by projecting a light source at the center of a globe onto a plane that is tangent to (touching) the globe's surface. The resulting increasingly severe distortion as distance increases from the standard point prevents showing a full hemisphere on one projection.

a) Where is the projection surface tangent to (touching) the globe? (See end flap of this lab manual.)

_____

_____

b) What kind of distortion, if any, occurs on a planar projection, and where?

_____

_____

c) Can you show the entire Earth on a single gnomonic projection? If not, why not?

_____

_____

d) Which areas of the globe would likely be mapped on a planar projection, and why? _____

_____

_____

The Mercator projection is useful in navigation and has become the standard for nautical charts since Gerardus Mercator, a Flemish cartographer, devised it for navigational purposes in 1569. The advantage of the Mercator projection is that lines of constant compass direction, called **rhumb lines**, are straight and thus facilitate the task of plotting compass directions between two points. However, these lines of constant direction, or bearing, are not the shortest distance between two points.

From the *planar* class of projections a gnomonic projection possesses a valuable feature: all straight lines are *great circles*—the shortest distances between two points on Earth's surface. A gnomonic projection is used to determine the coordinates of a great-circle (shortest) route; these coordinates are then transferred to a Mercator (true-direction) projection for determination of precise compass headings—the route along which the pilot or captain steers the airplane or ship.

Complete:

7. Use the two maps in Figure 3.5 to plot a great circle route between San Francisco (west coast of the United States, where you can see small details of San Francisco Bay) and London (southern England at the 0° prime meridian). The straight line on the gnomonic projection will show the shortest route between the two cities. Transfer the coordinates of this route over to the Mercator map, and connect with a line plot, to show the route's track. Note the route arching over southern Greenland on the Mercator.

8. Assume you board a plane in San Francisco for a flight to London. Briefly describe your great-circle flight route between the two cities (plotted on the map in Figure 3.5b). What are some of the features over which you fly? Describe the landscape and water below.

_____

_____

_____

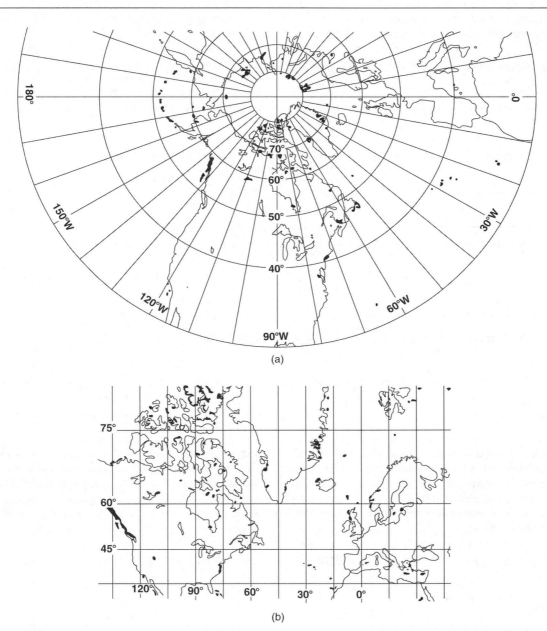

(a)

(b)

Figure 3.5
(a) Gnomonic/planar projection; (b) Mercator/cylindrical projection.

✳ SECTION 3

Map Scale

The ratio of the measured distance on a map to the actual distance on the ground is called **scale**; it relates a unit on the map to a similar unit on the ground. A 1:1 scale would mean that a centimeter on the map represents a centimeter on the ground, or an inch on the map represents an inch on the ground. A more appropriate scale for a local map is 1:24,000, in which 1 unit on the map represents 24,000 identical units on the ground.

Map scales may be presented in several ways. A *written* (or verbal) *scale* simply states the ratio using words—for example, "one centimeter to one kilometer" or "one inch to one mile." A *representative fraction* (RF or fractional scale) can be expressed with either a : or a /, as in 1:125,000 or 1/125,000. No actual units of measurement are mentioned because any unit is applicable as long as both parts of the fraction are in the same unit: 1 cm to 125,000 cm, 1 in. to 125,000 in., or even 1 arm length to 125,000 arm lengths, and so on. A *graphic* (bar) *scale* is a line or a bar divided into segments illustrating the correct map-to-ground ratio (Figure 3.6).

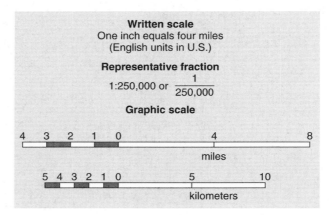

Figure 3.6
Three expressions of map scale: written scale, representative fraction, and graphic (bar) scale.

Scales are called "small," "medium," and "large," depending on the ratio described. Thus, a scale of 1:24,000 is a large scale, whereas a scale of 1:50,000,000 is a small scale. The greater the size of the denominator in a fractional scale (or number on the right in a ratio scale), the smaller the scale and the more abstract the map must be in relation to what is being mapped.

Examples of selected representative fractions and written scales are listed in Table 3.1 for small-, medium-, and large-scale maps.

Table 3.1

Sample representative fractions and written scales for small-, medium-, and large-scale maps.

System	Scale Size	Representative Fraction	Written Scale
English	Small	1:3,168,000	1 in. = 50 mi
		1:2,500,000	1 in. = 40 mi
		1:1,000,000	1 in. = 16 mi
		1:500,000	1 in. = 8 mi
		1:250,000	1 in. = 4 mi
	Medium	1:125,000	1 in. = 2 mi
		1:63,360 (or 1:62,500)	1 in. = 1 mi
		1:31,680	1 in. = 0.5 mi
		1:30,000	1 in. = 2500 ft
	Large	1:24,000	1 in. = 2000 ft
System		**Representative Fraction**	**Written Scale**
Metric		1:1,000,000	1 cm = 10.0 km
		1:50,000	1 cm = 0.50 km
		1:25,000	1 cm = 0.25 km
		1:20,000	1 cm = 0.20 km

Assume:

If a world globe is 61 cm (24 in.) in diameter and we know Earth has an equatorial diameter of 12,756 km (7926 mi), then the scale of the globe is the ratio of 61 cm to 12,756 km. Therefore, in order to determine the globe's representative fraction:

$$\text{RF} = \frac{\text{diameter of globe}}{\text{diameter of Earth}} = \frac{61 \text{ cm}}{12,756 \text{ km}} = \frac{61 \text{ cm}}{1,275,600,000 \text{ cm}} = \frac{1 \text{ cm}}{20,911,000 \text{ cm}}$$

Thus, the representative fraction for the 61 cm globe is expressed in centimeters as 1:20,900,000 (rounded off). This representative fraction can now be expressed in *any* unit of measure, metric or English, as long as both numbers are in the same units.

Complete:

1. Several companies manufacture large world globes for display in museums, corporate lobbies, and science exhibit halls. If such a globe featured a *diameter* of 4 m (13 ft), what is the scale of this globe expressed as a representative fraction? (Show your work.) (Hint: Earth's equatorial diameter = 12,756 km, [7926 miles])

_____

_____

2. There are 100,000 cm in 1 km (100 cm/m × 1000 m/km); the metric system needs going over in all of my classes. Just adding 1 km = 1000 m won't be enough. Given the large globe in question #1 and your calculation of its representative fraction, convert the RF to a *written scale* for this globe (1 cm on the globe = ? km on Earth). (Show your work.)

_____

_____

3. Use the topographic maps in the section at the back of the lab manual and a ruler (inches or cm) to measure the map distance between the points indicated. Noting the RF scales posted on each map, calculate the approximate ground distance (miles or km). Keep in mind that there are 63,360 inches in a mile: 12 in/ft × 5280 ft/mile. Compare results with the graphic scale. (Show your work.)

   a) **Omaha North Topographic Map #3:** length of the longest runway (NW-SE) at Eppley Airfield

   _____

   b) **Mt. Rainier National Park Topographic Map #6:** from Columbia Crest to McClure Rock

   _____

   c) **Ennis, Montana Topographic Map #9:** along the road from Lawton Ranch west to the highway junction

   _____

4. Find three different maps or globes as follows. Briefly describe the content of each and record the scale as a representative fraction or other scale noted on each map or globe.

   a) Classroom globe _____

   b) Map from your atlas _____

   c) Topo map at back of this manual _____

5. Give an example of a specific scale for each of the following:

   a) Small scale: _____

   b) Medium scale: _____

   c) Large scale: _____

6. What is an appropriate scale to use in preparing a world map for a classroom (pull-down "wall" map)? Why?

   _____

7. If you wanted a map to help you plan a local trip within your state or province, what scale would be best?

   _____

8. **Challenge question:** If you were helping to plan the painting of a world map on a local school playground that is approximately 10 meters along the equator, what scale would be best?

   RF: _____

   Written scale: _____

9. Using the Metric to English conversion chart on the inside of the fold-out back flap of this manual, what would be the approximate length of the equator in feet?

   _____

✱ SECTION 4

Township and Range Survey System

The Public Lands Survey System (1785) delineated locations in the United States. After modifications and adjustments, the final plan called for the establishment of the **township and range** grid system based on an initial survey point. For those portions of the United States surveyed under this system, the *initial points* are presented in Figure 3.7. Places are designated as north or south of the base line and east or west of the princi-

pal meridian. For example, Township 3 N means 3 "tiers" north of the base line and Range 4 W means 4 "columns" west of the principal meridian. Note that due to the latitudinal extent of California (southern border: 32°40′ N; northern border: 42° N), three initial points are used.

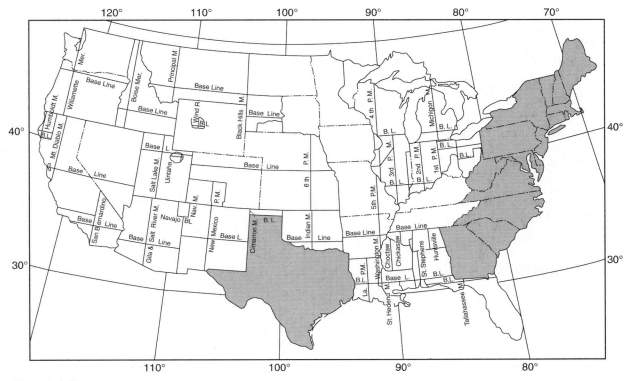

Figure 3.7
Principal meridians and base lines used for the township and range public-land surveys. Shaded states were not surveyed with the PLSS. (Adapted from the Bureau of Land Management, *Manual of Surveying Instructions*, Washington, DC, 1947, p. 168.)

The township and range grid system is based on 6-mile square *townships*, subdivided into 1-mile square *sections*, and quarter-section *homesteads*. Figure 3.8 portrays this system and shows the derivation of a parcel of land described as "S1/2, NW1/4, NE1/4, SW1/4,

Sec. 26, T2S, R4E." Note that you start from the smallest parcel of land ("the south half, of the northwest quarter, of the northeast quarter …"). Some land deeds still reflect this system.

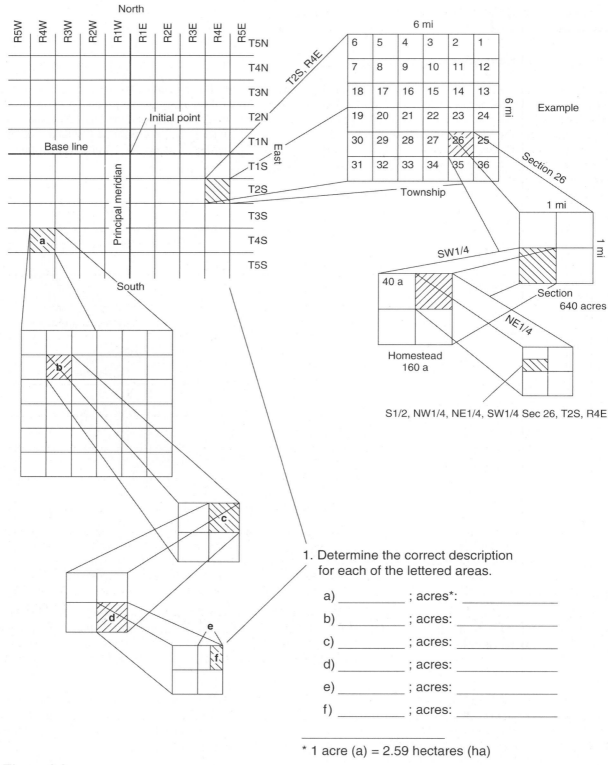

Figure 3.8

1. Determine the correct description for each of the lettered areas.

a) _____ ; acres*: _____

b) _____ ; acres: _____

c) _____ ; acres: _____

d) _____ ; acres: _____

e) _____ ; acres: _____

f) _____ ; acres: _____

_____

* 1 acre (a) = 2.59 hectares (ha)

Figure 3.8
Township and range rectangular survey grid system illustrating a parcel of land at: S1/2, NW1/4, NE1/4, SW1/4, Sec 26, T2S, R4E.

Questions and analysis about the township and range system.

1. Complete the fill-ins presented in Figure 3.8.

2. Using Figure 3.7 as your guide, which meridian and base line pair would be used to survey the area you are in?

_____

_____

3. If your campus was surveyed using the public-lands system, give a full description of the campus in standard township and range format (consult a topographic map for your campus area):

_____

_____

_____

_____

4. If the state or region where your campus is located was surveyed under a different system (example, parts of Texas, the original thirteen colonies), describe it relative to this system:

_____

_____

_____

_____

_____

Because meridians converge toward the North Pole, the linear distance between degrees of longitude decreases with higher latitude as illustrated in Lab Exercise 1. To maintain a more uniform rectangular shape for the townships, adjustments are made at intervals of every four townships to avoid the narrowing of range lines (Figure 3.9a).

On the maps on the next page note this adjustment between T15S and T16S and along the state line between Kansas and Missouri on the Olathe, Kansas—Missouri 1:100,000 quadrangle map shown in Figure 3.3b. You may have experienced these adjustments when driving along roads, especially in rural areas. The highway abruptly makes right-angle turns with warning markers to prevent you from ending up in a field.

5. Why are range lines adjusted every four townships? Relate your answer to Earth's coordinate grid system. Describe how the shape of a township in northern Alaska and a township in southern Texas would look if adjustments were not made.

_____

_____

_____

_____

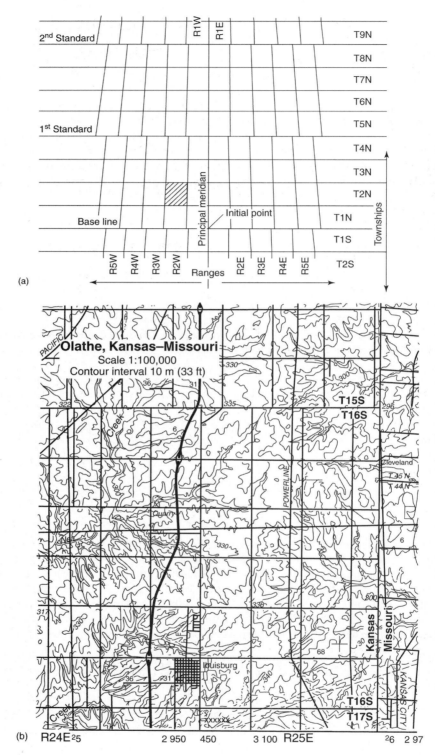

Figure 3.9

(a) North-south range lines narrow as meridians gradually converge toward the north. Secondary *standard parallels* are used to augment the primary *base line* and are added at intervals of every four townships to avoid a narrowing of ranges. (b) Range-line adjustments on the Olathe, Kansas-Missouri, 30′×60′ quad, 1:100,000 scale, 1983; prepared as part of the now-canceled USGS program to produce metric topographic maps.

Lab Exercise 4

Earth-Sun Relationships, Insolation, and Seasons

Earth's systems are powered by a constant flow of **insolation**—intercepted solar radiation. Earth intercepts only one two-billionth of the entire solar output, yet this is the singular significant source of energy for living systems. This exercise examines the nature of this energy and contrasts the solar output with energy reradiated by Earth's surface back to space. Sun's rays arrive at the top of the atmosphere in parallel beams, but the curvature of Earth presents surfaces at differing angles to this incoming radiation: lower latitudes receive more direct (vertical) illumination, whereas the higher latitudes receive more indirect (oblique)

rays. This pattern of uneven heating produces an energy imbalance in the atmosphere and on Earth's surface below—namely, equatorial and tropical energy surpluses and polar energy deficits.

Earth's systems are further influenced by shifting seasonal rhythms produced by changing daylength and Sun altitude (height of the noon Sun above the horizon at a given location). These seasonal changes become more noticeable toward the poles. Seasons occur as a result of Earth-Sun relationships produced by rotation, revolution, tilt of the axis, axial parallelism, and Earth's sphericity. Lab Exercise 4 has four sections.

Key Terms and Concepts

analemma
circle of illumination
daylength
declination
equinox
insolation

latitude
march of the seasons
revolution
rotation
solstice
subsolar point

Objectives

After completion of this lab, you should be able to:

1. *Label* diagrams representing Earth-Sun relations in the annual march of the seasons in order to *understand* the cause of seasons.
2. *Identify* the pattern of daily energy receipts at the top of the atmosphere at different latitudes throughout the year and *graph* the observations.
3. *Determine* the angle of incidence of insolation and the resultant intensity of the solar beam for various latitudes.
4. Utilize the analemma to *determine* the subsolar point, latitude, and noon Sun angle (altitude) for several locations using the analemma.

Materials/Sources Needed

pencil
calculator
color pencils
ruler

Lab Exercise and Activities

✳ SECTION 1

Earth-Sun Relations—Seasonality

Because Earth is spherical, its curved surface presents a continually varying angle to the incoming parallel rays of insolation. Differences in the angle of solar rays at each latitude result in an uneven distribution of insolation and heating. The place receiving maximum insolation is the point where insolation rays are perpendicular to the surface (radiating from directly overhead), called the **subsolar point.** All other places receive insolation at less than a 90° angle and thus experience more diffuse energy receipts.

Solar beam angles become more pronounced at higher latitudes. As a result, during a year's time, the top of the atmosphere above the equatorial region receives 2.5 times more insolation than that received above the poles.

1. On the accompanying diagrams in Figure 4.1, complete the following items. June has been started for you. Also, you may want to consult your physical geography text for reference.

   a) Extend the Sun's rays until they *intersect* (pass through) or are *tangent* to Earth's surface (touch at only one point). (See the Earth profile diagram for examples.)

   b) Where the rays intersect or are tangent to Earth's surface, draw a short line tangent to the surface, indicating the angle at which the Sun's rays are intercepted.

   c) Label the rays with the appropriate term—*vertical ray* (striking Earth at a 90° angle), *oblique ray* (striking Earth at less than a 90° angle), or *tangent ray* (parallel to Earth)—and mark and label the *subsolar point.*

   d) Draw the following on the Earth profiles and label:

   North and South Poles
   Equator
   Tropics of Cancer and Capricorn
   Circle of Illumination
   Arctic and Antarctic Circles

   e) Lightly shade the portion of Earth that is experiencing night, or the night half of the circle of illumination. Half of Earth is in sunlight and half is in darkness at any moment. The traveling boundary that divides daylight and darkness is called the **circle of illumination**—day–night dividing circle.

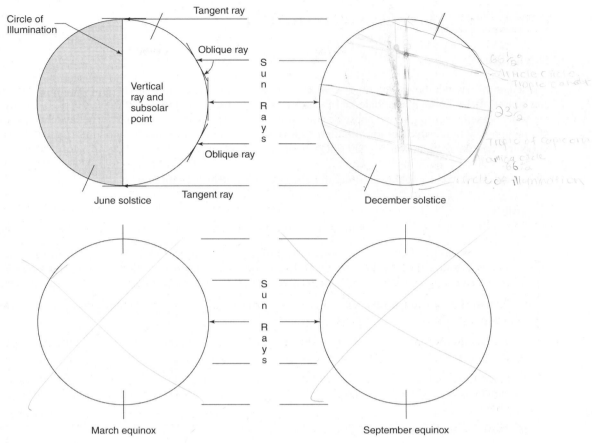

Figure 4.1
Earth-Sun relationships

2. Describe the changing position of the subsolar point and vertical rays throughout the four key seasonal positions of Earth.

_____

_____

_____

3. Describe the changing alignment of the circle of illumination in relationship to the poles throughout the four

key seasonal positions of Earth. _____

_____

_____

_____

4. What is the angular relationship between the subsolar point and the circle of illumination? _____

_____

❋ SECTION 2

Daylength

The length of daylight hours, the time between sunrise and sunset, varies both with latitude and the season of the year. Seasonal shifting of the subsolar point results in a realignment of the **circle of illumination** relative to the poles. See the diagrams in ❋ SECTION 1 of this exercise, and note the proportions of each latitude line (equator, Tropics of Cancer and Capricorn, Arctic and Antarctic Circles) that are in daylight vs. darkness at the various seasons.

1. Complete Table 4.1 (below), filling in the daylength at selected latitudes.

Table 4.1
Daylength—the time between sunrise and sunset—at selected latitudes for the Northern Hemisphere.

	Winter Solstice (December Solstice) December 21-22			Vernal Equinox (March Equinox) March 20-21			Summer Solstice (June Solstice) June 20-21			Autumnal Equinox (September Equinox) September 22-23		
	A.M.	P.M.	Daylength	A.M.	P.M.	Daylength	A.M.	P.M.	Daylength	A.M.	P.M.	Daylength
0°	6:00	6:00	_____	6:00	6:00	_____	6:00	6:00	_____	6:00	6:00	_____
30°	6:58	5:02	_____	6:00	6:00	_____	5:02	6:58	_____	6:00	6:00	_____
40°	7:26	4:34	_____	6:00	6:00	_____	4:34	7:26	_____	6:00	6:00	_____
50°	8:05	3:55	_____	6:00	6:00	_____	3:55	8:05	_____	6:00	6:00	_____
60°	9:15	2:45	_____	6:00	6:00	_____	2:45	9:15	_____	6:00	6:00	_____
90°	No sunlight			Rising Sun			Continuous sunlight			Setting Sun		

2. Explain how changes in the Sun's altitude (angle of incidence) and daylength (exposure) form the basis of seasonal change at these four locations throughout the year. _____

_____

3. Estimate the approximate length of daylight for the following locations:

   a) Dawson, Yukon Territory, Canada (64° N) on December 21 _____

   b) Adelaide, South Australia (35° S) on June 21 _____

   c) Bangkok, Thailand (14° N) on March 20 _____

   d) your location on June 21 _____

Distribution of Insolation at the Top of the Atmosphere

As we just learned, the spherical Earth presents a curved surface to the Sun's parallel rays, producing an uneven distribution of energy across the latitudes. Only the subsolar point receives insolation from directly overhead: all other locations receive Sun's rays at increasingly lower angles of incidence as their distance increases from the subsolar point. A lower angle of insolation is less intense at the surface, that is, the energy arriving is more diffuse. Sun's rays striking a surface at a 30° angle are diffused over twice the surface area as that receiving a perpendicular beam. Figure 4.2 shows that these oblique rays, striking Earth's surface at a 30° angle at point B, cover twice as much surface area (and are thus more diffuse and only one-half as intense) as the direct rays, which strike at a 90° angle at the subsolar point A.

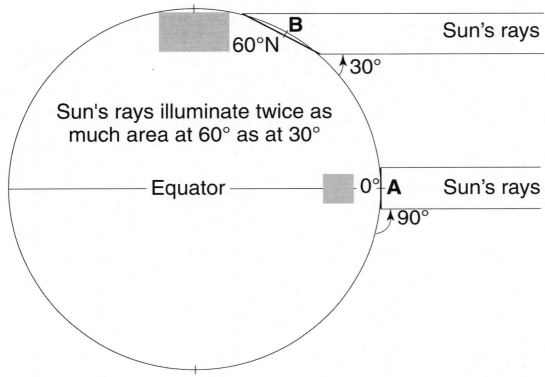

Figure 4.2
Insolation angles and concentration of Sun's energy

In addition, the greater length of travel through the atmosphere at higher latitudes reduces solar intensity due to increased reflection, absorption, and scattering within the atmosphere. This energy receipt also varies daily, seasonally, and annually as Sun angle and daylength vary—less toward the equator, more toward the poles.

A useful altitude at which to characterize insolation is the top of the atmosphere (480 km, 300 mi). The graph in Figure 4.3 plots the daily variation in insolation for selected latitudes. Latitudes are marked along the left side in 10° intervals. The months of the year are marked and labeled across the top and bottom. The dashed curved line shows the declination of the Sun (the latitude of the subsolar point) throughout the year.

To read the graph, select a latitude line and follow it from left to right through the months of the year. The daily insolation totals are read from the curved lines, given in watts per square meter per day (W/m^2 per day). For example, at the top of the atmosphere above 30° N latitude the insolation values (in W/m^2 per day) are as follows (readings are approximate—extrapolate when necessary): 240 on January 1; 250 on January 15, and 275 on February 1.

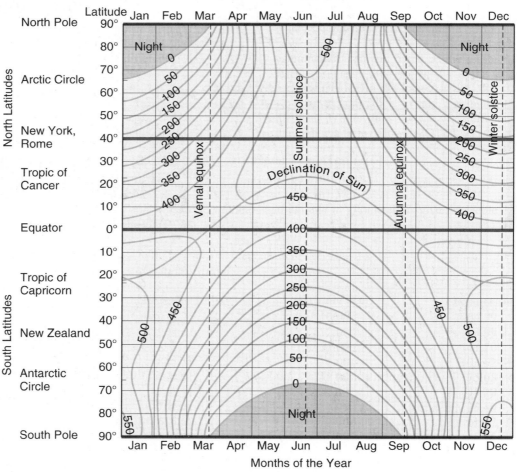

Figure 4.3

Total daily insolation received at the top of the atmosphere charted in watts per square meter by latitude and month. (1 watt/m^2 = 2.064 cal/cm^2/day) Adapted from the Smithsonian Institution Press from *Smithsonian Miscellaneous Collections: Smithsonian Meteorological Tables*, vol. 114, 6th Edition. Robert List, ed. Smithsonian Institution, Washington, DC, 1984, p. 419, Table 134.

1. Using the data in Figure 4.3 and the graph provided below, plot data for specific latitudes. The graph you create will enable you to note the changes throughout the year in the amount of insolation received at a given latitude. Compare the differences in annual insolation patterns at various latitudes. Plot all seven latitudes on one graph, using color pencils to distinguish each. (Be sure to include a legend.)

   • North Pole (started on the graph)
   • New York (started on the graph; city at 40° N)
   • Tropic of Cancer
   • Equator
   • South Pole

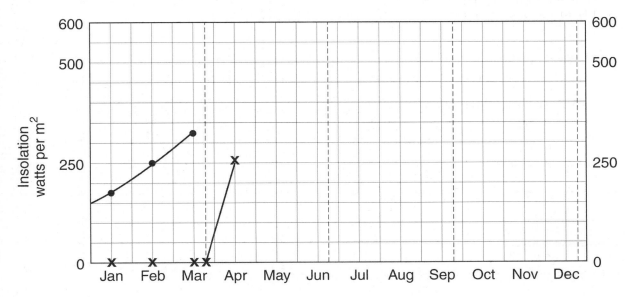

After completing the graph of the energy receipt at these different latitudes, complete the following:

2. Compare and contrast the plotted insolation values at the equator and either of the poles. What factors explain these different patterns of energy receipt?

   _____

   _____

   _____

3. What factor results in the June (summer) solstice energy receipt at the North Pole exceeding that received on the same day at the equator?

   _____

   _____

   _____

4. What is happening at the poles on the March (vernal) equinox? _____

_____

_____

5. What kind of generalization can you make about the relationship between latitude and annual variation in insolation? _____

_____

_____

6. Challenge question: Why do you think the South Pole receives over 550 W/m^2 at the December solstice whereas the North Pole receives over 500 W/m^2 during the June solstice? (There is a reason related to our annual relationship to the Sun.) _____

_____

_____

✳ SECTION 4

Seasonal Variation in the Sun's Declination and Altitude

Seasonal variations are a response to changes in the Sun's **altitude**, the angle between the horizon and the noon Sun. The Sun's **declination**, the latitude of the subsolar point, migrates annually through 47° of latitude—between the *Tropic of Cancer* at 23.5° N and the *Tropic of Capricorn* at 23.5° S. The **analemma** is a convenient device to track the passage of the Sun's path and declination throughout the year (Figure 4.4). The horizontal lines are latitudes from 25° N to 25° S. The "figure 8" is a calendar with each day of the year represented by a black or white segment.

Find August 20th on the analemma; note that it lies on the 12° N line—the subsolar point on that date. While you should remember the subsolar points for the solstices and equinoxes, an analemma will help you determine the declination for the other days of the year.

The analemma also is useful for ascertaining the positive and negative equations of time and periods of "fast-Sun times" (greatest in October and November) and "slow-Sun times" (greatest in February and March). A 24-hour (86,400 seconds) average day determines *mean solar time*. However, an *apparent solar day* is based on observed successive passages of the Sun over a given meridian. Any difference between observed solar time and mean solar time is called the *equation of time*. On successive days, if the Sun arrives overhead at a meridian *after* 12:00 noon local standard time (taking longer than the 24 hours of a mean solar day)—like an airliner arriving later than scheduled—the equation of time is *negative*, and the Sun is described as "slow." If, on the other hand, the Sun arrives overhead *before* 12:00 noon local time on successive days (taking less time than 24 hours)—like an airliner arriving ahead of schedule—the equation of time is *positive*, and the Sun then is described as "fast." The combination of the tilt and the eccentricity of Earth's orbit around the sun produces this difference between mean solar time and apparent solar time. Because Earth's orbit is elliptical rather than circular, Earth actually travels more quickly when we are at perihelion and more slowly when we are at aphelion. The solar day increases in length by up to 7.9 seconds for several months and then decreases in length for several months. This difference accumulates from day to day and results in the equation of time ranging from up to 16 minutes slow and 14 minutes fast at differing times during the year. The equation of time is across the top of the analemma.

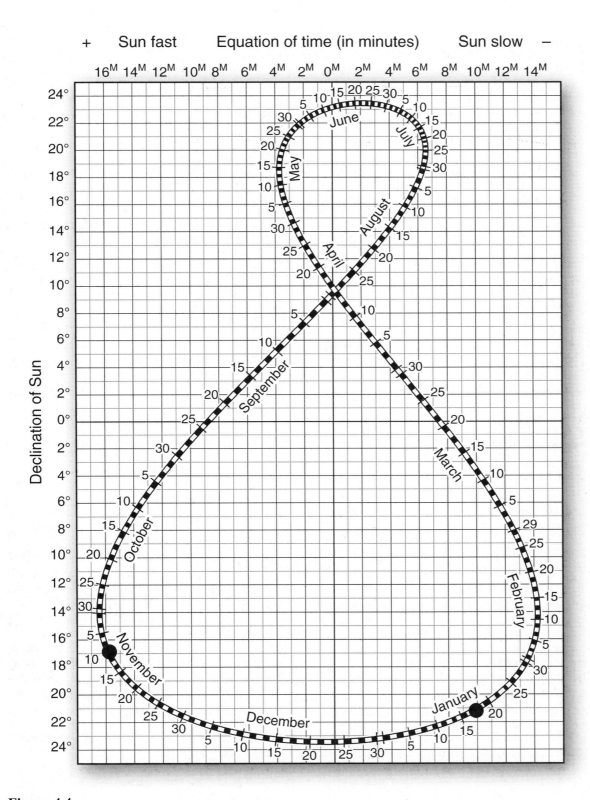

Figure 4.4
The analemma

Find January 17th on the analemma (marked with a dot). It lies on the line labeled 9 minutes, on the "Sun slow" side. This means that the Sun will reach its "noon" *zenith* (highest point in the sky on that day) at 12:09 P.M., 9 minutes "late."

1. Use the analemma to find:

   a) the subsolar point on November 10 (marked) _____ [*17° south latitude*] _____

   b) the subsolar point on May 11 _____

   c) the subsolar point today _____

   d) the date(s) when the declination is 9° S _____

   e) the date(s) when the declination is 21° N _____

2. At what clock time does the Sun actually reach zenith on:

   October 13? 11:47 A.M _____ [*The sun is 13 minutes fast so it will reach its zenith at 11:47 A.M.*] __

   March 8? _____

   May 20? _____

   Today? _____

As we have noted, the noon Sun angle changes at any given latitude throughout the year. Using the analemma to determine the subsolar point (Sun's declination), you can calculate the altitude of the noon Sun (Sun $\angle$ for any location by using the following formula:

$$\angle = 90° - (\text{arc distance between your latitude and subsolar point})$$

EXAMPLES:

You are at 30° N on December 21. You know that on the December solstice the subsolar point of the noon Sun is 23.5° S. Using the above formula:

$\angle = 90° - (\text{arc distance between your latitude and subsolar point})$

$\angle = 90° - (30° \text{ N} \leftrightarrow 23.5° \text{ S})$(*the symbol* $\leftrightarrow$ *is to indicate* $\leftrightarrow$ *"arc distance between"*)

$\angle = 90° - 53.5°$

$\angle = 36.5°$ above the southern horizon from your location

You determined that on December 21, if you were standing at 30° north latitude, you observe the Sun's noon altitude at 36.5° above the southern horizon (since you are north of the subsolar point on this day).

If you are at 40° N on April 20, you determine from the analemma that the subsolar point is 10° N; therefore:

$\angle = 90° - (\text{arc distance between your latitude and subsolar point})$

$\angle = 90° - (40° \text{ N} \leftrightarrow 10° \text{ N})$

$\angle = 90° - 30°$

$\angle = 60°$ above the southern horizon *

* Note: You would observe the same Sun altitude on August 25 when the subsolar point is once again at 10° N. If you end up with a negative value for our Sun altitude, then the Sun is below the horizon and will not be visible on that day—as someone might experience north of the Arctic Circle in the Northern Hemisphere winter.

3. You are vacationing at Disney World near Orlando, Florida (28.5° N) on June 3. At what altitude will you observe the noon Sun? Include correct horizon. (Show work.)

$\angle = 90° - (\text{arc distance between your latitude and subsolar point})$

$\angle = 90° - (28.5° \text{ N} \leftrightarrow \text{subsolar point})$

_____

_____

4. On that same day friends of yours are sightseeing at Iguaçu Falls in Brazil (25.7° S); at what altitude will they observe the noon Sun?

_____

_____

_____

5. Calculate the altitude of the noon Sun if you are vacationing in Kuala Lumpur, Malaysia (3.2° N) on July 25.

_____

_____

_____

6. Suppose you were going to install PV solar panels to generate electricity from the Sun at your house and you wanted the Sun to shine on them at a 90° angle on the equinoxes. What angle would they need to be based on the latitude of your home? Should they face north or south?

_____

_____

_____

7. Using the graphs provided, plot the seasonal change in the Sun's noon altitude for:

a) North Pole (90° N)

b) La Ronge, Saskatchewan (55° N)—completed for you

c) Atlanta, Georgia (34° N)

d) Dunedin, New Zealand (46° S).

Complete these calculations only for each of the four seasonal anniversary dates at the solstices and equinoxes (March 20, June 20, September 22, and December 21). If needed, refer to your geography text to determine the Sun's declination on these dates for each of the four locations. Then, use the procedure above to determine the Sun's altitude in the sky at noon that you would observe if you were standing at each location.

Whether you view the noon Sun above your northern or your southern horizon depends upon your latitude. The subsolar point is limited to the latitudinal belt between the Tropics of Cancer and Capricorn. An observer who is poleward of the tropics will see the noon Sun only above the "equatorward" horizon (above the southern horizon if north of the Tropic of Cancer, and above the northern horizon if south of the Tropic of Capricorn). An observer on or between the tropics will measure the noon Sun above the northern horizon part of the year, above the southern horizon part of the year, and directly overhead on 1 or 2 days.

Use the analemma in Figure 4.4 to help you answer the following questions:

8. a) If you are at 5° N latitude, on what dates will the noon Sun be directly overhead?

_____

b) Approximately how many days will you view the noon Sun above your northern horizon? _____

… above your southern horizon? _____

c) If you are at 18° S latitude, on what dates will the noon Sun be directly overhead?

_____

d) Approximately how many days will you view the noon Sun above your northern horizon? _____

… above your southern horizon? _____

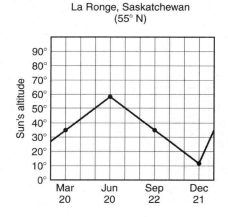

North Pole
(90° N)

La Ronge, Saskatchewan
(55° N)

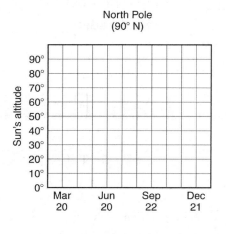

Atlanta, Georgia
(34° N)

Dunedin, New Zealand
(46° S)

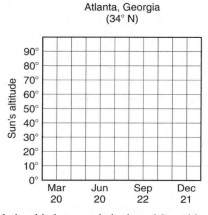

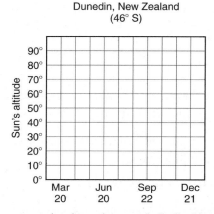

Using the same relationship between latitude and Sun altitude, you can also determine your latitudinal location by measuring the solar noon altitude, which can be done with a sextant, or more simply with a protractor and a stick. For example:

Using a sextant, you measure the noon Sun angle at 59° above the southern horizon on March 11. From the analemma you know that the subsolar point is 4° S; therefore:

$$\angle = 90° - (\text{arc distance between your latitude and subsolar point})$$

$$59° = 90° - (\text{your latitude} \leftrightarrow 4° \text{ S})$$

$$59° = 90° - 31°$$

If the arc distance between your latitude and the subsolar point = 31°, and the subsolar point is 4° S, then your latitude must be 27° N (31° − 4°). (You must be north latitude because the Sun is viewed above your southern horizon.)

This method of determining latitude, used along with chronometers and local time to determine longitude (Lab Exercise 1), allowed sailors or other navigators to obtain their absolute global location in the days before GPS (Global Positioning System) units were devised.

9. Calculate the latitude of a mariner who has measured the noon Sun altitude as 60° above the southern horizon

   on August 13. _____

   _____

10. What would the mariner's latitude be if the Sun had been measured above the northern horizon? _____

   _____

Name: _____   **Laboratory Section:** _____

Date: _____   **Score/Grade:** _____

Lab Exercise 5

Temperature Concepts

Temperature, a measure of sensible heat energy present in the atmosphere and other media, indicates the average kinetic energy of individual molecules within the atmosphere. You have observed air temperature as part of the "Weather Calendar" exercise in the Prologue Lab.

A variety of temperature regimes worldwide affect cultures, decision making, and resources consumption. Global temperature patterns apparently are changing in a warming trend and are the subject of much scientific, geographic, and political interest. In Lab Exercise 5, we examine some temperature concepts and graph the temperature profile of the atmosphere. The effect of temperature in combination with wind and humidity is also presented, since wind chill and the heat index produce apparent temperatures that represent a risk in regions of North America. Lab Exercise 5 features five sections.

Key Terms and Concepts

apparent temperature (sensible temperature)
heat index
kinetic energy
mesosphere
sensible heat

stratosphere
temperature
thermosphere
troposphere
wind chill factor

Objectives

After completion of this lab, you should be able to:

1. *Graph* altitudinal temperature changes in the atmosphere.
2. *Differentiate* between and convert metric and English units of temperature measures.
3. *Graph* and *analyze* global temperature anomalies and carbon dioxide levels.
4. *Obtain* temperature data over a three-day period.
5. *Determine* wind-chill factors for various temperature and wind situations, and heat index factors for various temperature and humidity situations.

Materials/Sources Needed

pencil
color pencils
highlighter pen
calculator

Lab Exercise and Activities

✳ SECTION 1

Temperature Profile of the Atmosphere

Based on *temperature*, the atmosphere is divided into four distinct zones or layers: the **thermosphere, mesosphere, stratosphere,** and **troposphere.** The transition area at the top of each temperature region is named using the suffix *-pause,* which means "to cause to change," i.e., **thermopause, mesopause, stratopause,** and **tropopause.**

Temperatures in the lower atmosphere do not simply decline with altitude, for it is more complex than this and actually varies. In the upper atmosphere the temperature profile shows that temperatures rise sharply in the thermosphere, up to 1200°C (2200°F) and higher. However, we must use different concepts of "temperature" and "heat" to understand this effect.

The intense radiation in this portion of the atmosphere excites individual molecules (nitrogen and oxygen) and atoms (oxygen) to high levels of vibration. This **kinetic energy,** the energy of motion, is the vibrational energy stated as "temperature." However, the density of the molecules is so low that little actual heat is produced. Heating in the lower atmosphere near Earth's surface differs because the greater number of molecules in the denser atmosphere transmit their kinetic energy as **sensible heat,** meaning that we can sense it and measure it. Figure 5.1 gives you the general trend for this temperature profile that you specifically plot in this section using data from Table 5.1.

Table 5.1
Standard temperature values for the atmosphere

Layer of Atmosphere	Altitude	Temperature
Surface (No. Hemis.)	Sea level	15°C (59°F)
Tropopause	18 km (11 mi)	−57°C (−70°F)
[Isothermal layer]	[11 to 25 km]	
Stratopause	50 km (31 mi)	0°C (32°F)
Mesopause	80 km (50 mi)	−90°C (−130°F)
Thermopause	480 km (300 mi)	1200°C (2200°F)+

1. Using the graph in Figure 5.1, plot the standard temperature values given in Table 5.1 (the sea level value has been done for you). After you plot the data points, connect them with a line graph to complete the profile. Label the layers of the atmosphere and the transition areas at the top of each layer.

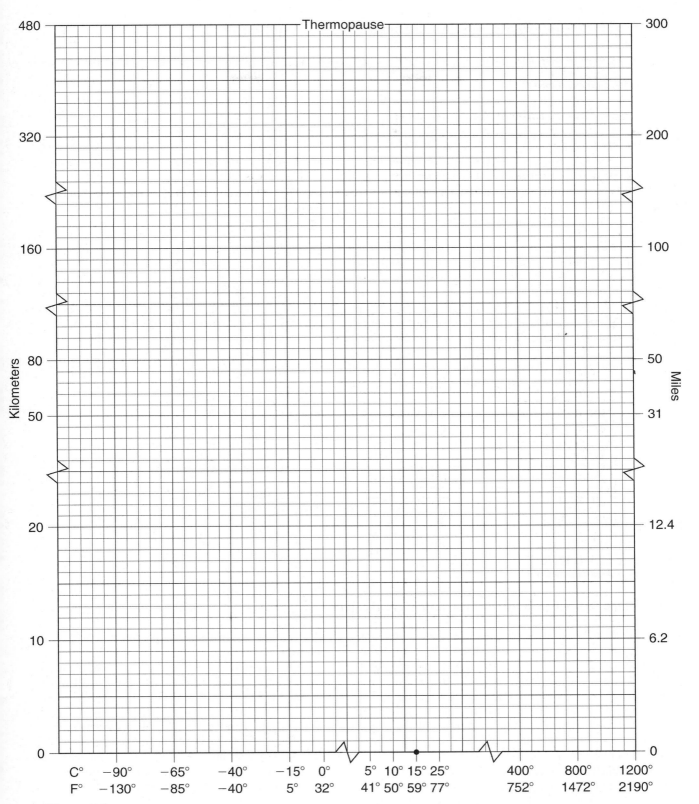

Figure 5.1
Atmospheric temperature profile graph

Analysis and completion questions about the standard temperature profile.

2. Briefly explain why the temperature decreases as altitude increases in the troposphere (at the normal lapse

rate). _____

_____

_____

3. Why do temperatures increase throughout most of the stratosphere? Specifically discuss the process that

produces this warming effect. _____

_____

_____

4. After consulting the appropriate sections of a physical geography textbook, <u>briefly</u> describe the predicament
relative to stratospheric ozone: identification and history of the problem, causes, current actions being taken,

and present status. _____

_____

_____

_____

_____

✳ SECTION 2

Temperature Concepts, Terms, and Measurements

The amount of heat energy present in any substance is measured and expressed as its temperature. Temperature actually is a reference to the speed at which atoms and molecules that make up a substance are moving. Changes in temperature are caused by the absorption (gain) or emission (loss) of energy. The temperature at which all motion in a substance stops is called 0° absolute temperature. Its equivalent in different temperature-measuring schemes is −273° Celsius (C), −459.4° Fahrenheit (F), or 0 Kelvin (K).

The Celsius scale (formerly called centigrade) is named after Swedish astronomer Anders Celsius (1701–1744). It divides the difference between the freezing and boiling temperatures of water into 100 degrees on a decimal scale, with freezing at 0° and boiling at 100°.

The Fahrenheit scale places the freezing point of water at 32°F (0°C, 273°K) and boiling point of water at 212°F (100°C, 373°K), with 180F° difference between freezing and boiling. The United States remains the only major country still using the Fahrenheit scale, named after the German physicist Daniel Fahrenheit (1686–1736) who invented the alcohol and mercury thermometers and who developed this thermo-metric scale in the early 1700s. Most countries use the Celsius scale to express temperature.

The Kelvin scale, named after British physicist Lord Kelvin (born William Thomson, 1824–1907) who proposed the scale in 1848, is used in science because temperature readings are proportional to the actual kinetic energy in a material.

The essential metric-English conversion formulas for temperature (°C or °F) are:

To convert °C to °F:

$$°F = (°C \times 1.8) + 32$$

$$\text{or} \quad °F = 1.8°C + 32$$

To convert °F to °C:

$$°C = (°F - 32) \div 1.8$$

$$\text{or} \quad °C = (°F - 32) \times 0.556$$

When recording air temperature, the degree symbol is placed before the C or F and is read *degree Celsius* (°C) or *degree Fahrenheit* (°F)—as in "yesterday was 20°C and today is 25°C." However, when expressing the difference between two temperatures, dealing with lapse rates, temperature ranges, or units of temperature, the degree symbol is placed after the C or F and is read *Celsius degree* (C°) or *Fahrenheit degree* (F°)—as in "yesterday was 5 C° cooler than today." The following metric-English conversions apply:

The essential metric-English conversion formulas for temperature (C° or F°) are:

To convert C° to F°:

$$F° = C° \times 1.8 \quad \text{or} \quad F° = 1.8\,C°$$

To convert F° to C°:

$$C° = F° \div 1.8 \quad \text{or} \quad C° = F° \times 0.556$$

---

1. Using the given conversion formulas, complete the following conversions. Note the placement of the degree symbol in your conversions.

   25°C _[77°F]*_   30°F _____   30 C° _____   41 F° _____

   4°C _____   92°F _____   64 C° _____   57 F° _____

   *(°F = ((25° × 1.8) + 32°) = (45° + 32°) = 77° F)

2. What temperature is it as you work on this lab exercise?

   a) outdoor temperature?   _____ °C   _____ °F

   b) indoor temperature?   _____ °C   _____ °F

3. Using a physical geography text, atlas, encyclopedia, or Internet, answer the following (*Geosystems*, eighth edition, Figure 5.2; *Elemental Geosystems*, sixth edition, Figure 3.15.). Be sure to list temperature record, date, and place.

   a) Highest natural temperature recorded on Earth (where and value)? _____

   _____

   b) Lowest natural temperature recorded on Earth (where and value)? _____

   _____

   c) Lowest natural temperature recorded for the Northern Hemisphere? _____

   _____

   d) Highest natural temperature recorded in North America? _____

   _____

Optional: Both *Geosystems* textbooks discuss the factors that produce such temperature readings (air pressure patterns, air masses, geographical location, etc.). See Chapters 5, 6, and 8 of *Geosystems* or Chapters 3, 4, and 5 of *Elemental Geosystems*.

    e)   What factors produce such temperature readings in North America (air pressure patterns, air masses,

        geographical location, etc.)? _____

_____

_____

✳ SECTION 3

Carbon Dioxide and Temperature

Carbon dioxide is a natural by-product of life processes and it is an important component of the atmosphere. Although it accounts for only a small percent of the atmosphere, about 0.038%, it plays a vital role in maintaining habitable global temperatures. Without CO_2 in the atmosphere, the temperature of Earth would be 15 C° cooler than it is, with Earth's average temperature around the freezing point of water. As carbon dioxide levels have increased in the atmosphere, so have global temperatures.

The Dome C ice core shows that present levels of carbon dioxide, methane, and nitrous oxides, and resulting air temperatures, are highest in the present decade than in the entire 800,000-year record! These are remarkable times in which we live. 2010 has tied 2005 as the warmest year ever recorded, 2000-2009 is the warmest decade ever recorded, and ten of the eleven warmest years ever recorded have occurred since 2000, beating records that stretch back to 1850.

The Intergovernmental Panel on Climate Change (IPCC), through major reports in 1990, 1992, 1995, 2001 and its *Fourth Assessment Report (AR4)* in 2007, confirms that global warming is occurring. The IPCC AR4 reached a consensus, in part stating:

> Warming of the climate system is unequivocal, as is now evident from observations of increases in global average air and ocean temperatures, widespread melting of snow and ice, and rising global average sea level.... Most of the observed increase in globally averaged temperatures since the mid-20th century is *very likely [greater than >90% probability]* due to the observed increases in anthropogenic greenhouse gas concentrations....*

The CO_2 data for this exercise comes from the Mauna Loa observatory on the big island of Hawai'i, and is the longest continuous direct measurement of CO_2. The project was begun by Dr. Keeling in 1958. When completed, the graph below will compare CO_2 levels with global temperature anomalies from 1958 until 2010. Pay close attention to the scales of the two data series since they are in different units. The temperatures are given as degrees Celsius warmer or cooler than the 1951–1980 mean values. The CO_2 units are parts per million (PPM), while the temperature values are in Celsius degrees.

1. Using the graph in Figure 5.2, plot the values given in Table 5.2 using blue dots for the CO_2 levels and red dots for the temperature values (the first several years of CO_2 and temperature have been done for you). After you plot the data points, connect them to make a line graph.

---

*Intergovernmental Panel on Climate Change (IPCC) *Fourth Assessment Report* (2007), Working Group I, *The Physical Science Basis*, "Summary for Policy Makers," (February 5, 2007), IPCC Secretariat, Geneva, Switzerland, pp. 5, 10. See: http://www.ipcc.ch/ to download the three Working Group summaries for free.

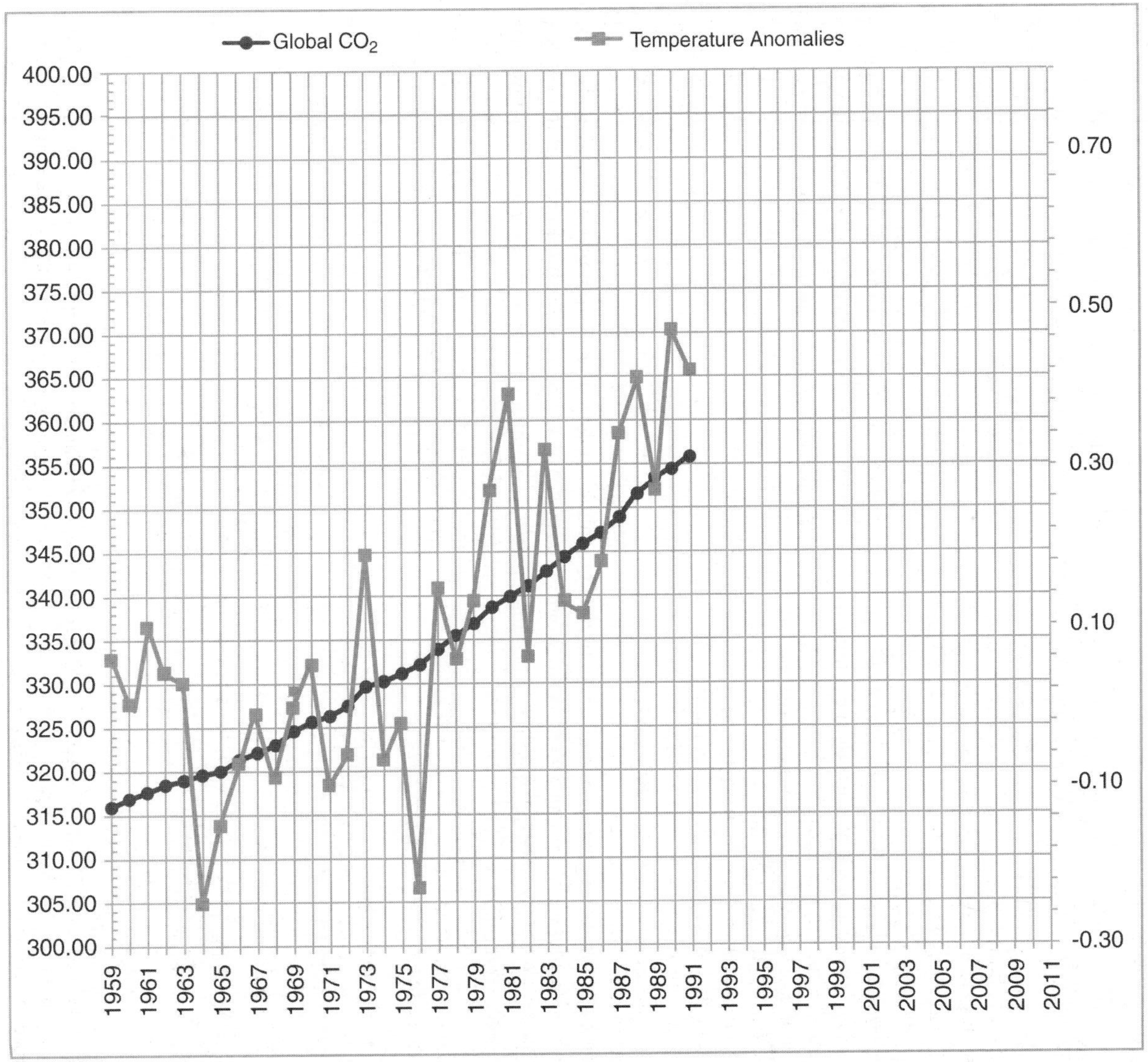

CO_2	Year	Temp Anom	CO_2	Year	Temp Anom
315.98	1959	0.05	347.19	1986	0.18
316.91	1960	-0.01	348.98	1987	0.34
317.64	1961	0.09	351.45	1988	0.41
318.45	1962	0.04	352.90	1989	0.27
318.99	1963	0.03	354.16	1990	0.47
319.62	1964	-0.25	355.48	1991	0.42
320.04	1965	-0.15	356.27	1992	0.14
321.38	1966	-0.07	356.95	1993	0.17
322.16	1967	-0.01	358.64	1994	0.30
323.04	1968	-0.09	360.62	1995	0.44
324.62	1969	0.00	362.36	1996	0.36
325.68	1970	0.05	363.47	1997	0.38
326.32	1971	-0.10	366.50	1998	0.70
327.45	1972	-0.06	368.14	1999	0.43
329.68	1973	0.19	369.40	2000	0.41
330.17	1974	-0.07	371.07	2001	0.56
331.08	1975	-0.02	373.17	2002	0.68
332.05	1976	-0.23	375.78	2003	0.65
333.78	1977	0.15	377.52	2004	0.59
335.41	1978	0.06	379.76	2005	0.77
336.78	1979	0.13	381.85	2006	0.65
338.68	1980	0.27	383.71	2007	0.73
340.11	1981	0.39	385.57	2008	0.55
341.22	1982	0.06	387.35	2009	0.72
342.84	1983	0.32	389.78	2010	0.83
344.41	1984	0.13		2011	
345.87	1985	0.12		2012	

Table 5.2
CO_2 and temperature values

Figure 5.2
Global CO_2 and temperature anomalies

2. After completing the CO_2 graph, calculate the increase in CO_2 over the past ten years, both as ppm per year and percent per year. How many ppm higher would levels be in 50 years, if they continue to increase at the present rate? What would the total ppm of CO_2 be at that time?

_____

_____

3. Calculate the average temperature anomaly for 2005–2009, 2000–2004, and 1996–1999. What is the trend in temperature anomalies over the past fifteen years? If this trend continues, what would the predicted temperature anomaly be in 50 years?

_____

_____

4. Looking at the graph, what do you notice about CO_2 and temperature levels immediately following the eruption of Mt. Pinatubo in 1991? Did the levels go up or down? By how much?

_____

_____

5. The values for CO_2 vary during the year with the highest values in March and lowest values in September. The average seasonal variation for CO_2 during this time period is 5.82 ppm. What would cause this to happen? What has been the average annual increase for CO_2 from 2001–2006? How does this amount compare with the seasonal variation?

_____

_____

6. The Kyoto Protocol calls for an overall reduction of greenhouse gases of 5.3% below the 1990 levels. What was the average level of CO_2 in 1990? What would the level of CO_2 be if the goals were met? How many ppm below current levels would that be? The "Group of 77" countries favor a 15% reduction below 1990 levels. What would that level be and how far below present levels is it?

_____

_____

✳ SECTION 4

Temperature Readings

Relative to atmospheric temperature, locate a properly installed thermometer either at the college or university you are attending, or perhaps at home. If you do not have access to a thermometer, find another source of information for local temperature: a radio or television station, a local cable channel, the Weather Channel, a local newspaper, the National Weather Service, Environment Canada, or one of the several Internet sources available.

Perhaps you have already done this in working on your "Weather Calendar" in the Prologue Lab. The purpose of this section is to identify a reliable source of temperature information. Long after this lab course is completed such information will benefit your lifestyle and activities. Also, this information is important to energy management of interior spaces.

1. Record the air temperature for three days at approximately the same time of day, if possible. For Lab Exercise 7 ✳ SECTION 2, also record the barometric (air pressure) reading and record these on the chart in Lab Exercise 7. See if you can detect a trend between air temperature and other atmospheric phenomena.

Temperature	Place of observation	Time of day; date
Day 1:	Place:	Time:
Day 2:	Place:	Time:
Day 3:	Place:	Time:

2. Weather conditions and state of the sky at the time of your observation:

Day 1: _____

_____

Day 2: _____

_____

Day 3: _____

_____

3. Do you have an outdoor thermometer available where you live? Describe the location of the thermometer from which you obtained these readings (orientation, exposure relative to compass direction, height from the ground). _____

_____

_____

4. Is temperature data available through the Geography Department at your school, or elsewhere on campus? Is there a campus weather station? _____

_____

✳ SECTION 5

The Temperatures We Feel

Our perception of temperature is described by the terms **apparent temperature**, or *sensible temperature*. The combination of water vapor content, wind speed, and air temperature affect each individual's sense of comfort. High temperatures, high humidity, and low winds produce the most heat discomfort, whereas low humidity and strong winds enhance cooling sensation and effect (due to the increased evaporation of moisture from the skin). The wind chill index is important to those who experience (live through) winters with freezing temperatures. The NWS and the Meteorological Services of Canada (MSC, **http://www.msc-smc.ec.gc.ca/contents_e.html/**) revised the wind chill formula and standard assumptions and introduced a new wind-chill chart during the 2001–2002 winter season. The new Wind Chill Temperature (WCT) Index is an effort to improve the accuracy of heat loss calculations.

A wind-chill chart of estimated values is presented in Figure 5.3 below in both metric and English units. The **wind chill factor** indicates the enhanced rate at which body heat is lost to the air. As wind speeds increase, heat loss from the skin increases. For example, if the air temperature is −1°C (30°F) and the wind is blowing at 32 kmph (20 mph), skin temperatures will be −8°C (17°F). The **heat index** (HI) indicates the human body's reaction to air temperature and water vapor. The water vapor in air is expressed as relative humidity, a concept presented in Chapter 7. For now, we simply can assume that the amount of water vapor in the air affects the evaporation rate of perspiration from the skin because the more water vapor in the air (the higher the humidity), the less water from perspiration the air can absorb through evaporation, thus reducing natural evaporative cooling. The heat index indicates how the air feels to an average person—in other words, its apparent temperature. Figure 6.8 is an abbreviated version of the heat index used by the NWS and now included in its daily weather summaries during appropriate months. The table beneath the graph describes the effects of heat-index categories on higher-risk groups. A combination of high temperature and high humidity can severely reduce the body's natural ability to regulate internal temperature. The NWS provides a heat-index forecast warning map (see http://www.nws.noaa.gov/om/heat/index.shtml); a heat-index calculator is at http://www.crh.noaa.gov/jkl/?n = heat_index_calculator.

Actual Air Temperature in °C (°F)

Calm	4° (40°)	−1° (30°)	−7° (20°)	−12° (10°)	−18° (0°)	−23° (−10°)	−29° (−20°)	−34° (−30°)	−40° (−40°)
8 (5)	2° (36°)	−4° (25°)	−11° (13°)	−17° (1°)	−24° (−11°)	−30° (−22°)	−37° (−34°)	−43° (−46°)	−49° (−57°)
16 (10)	1° (34°)	−6° (21°)	−13° (9°)	−20° (−4°)	−27° (−16°)	−33° (−28°)	−41° (−41°)	−47° (−53°)	−54° (−66°)
24 (15)	0° (32°)	−7° (19°)	−14° (6°)	−22° (−7°)	−28° (−19°)	−36° (−32°)	−43° (−45°)	−50° (−58°)	−57° (−71°)
32 (20)	−1° (30°)	−8° (17°)	−16° (4°)	−23° (−9°)	−30° (−22°)	−37° (−35°)	−44° (−48°)	−52° (−61°)	−59° (−74°)
40 (25)	−2° (29°)	−9° (16°)	−16° (3°)	−24° (−11°)	−31° (−24°)	−38° (−37°)	−46° (−51°)	−53° (−64°)	−61° (−78°)
48 (30)	−2° (28°)	−9° (15°)	−17° (−1°)	−24° (−12°)	−32° (−26°)	−39° (−39°)	−47° (−53°)	−55° (−67°)	−62° (−80°)
56 (35)	−2° (28°)	−10° (14°)	−18° (0°)	−26° (−14°)	−33° (−27°)	−41° (−41°)	−48° (−55°)	−56° (−69°)	−63° (−82°)
64 (40)	−3° (27°)	−11° (13°)	−18° (−1°)	−26° (−15°)	−34° (−29°)	−42° (−43°)	−49° (−57°)	−57° (−71°)	−64° (−84°)
72 (45)	−3° (26°)	−11° (12°)	−19° (−2°)	−27° (−16°)	−34° (−30°)	−42° (−44°)	−50° (−58°)	−58° (−72°)	−66° (−86°)
80 (50)	−3° (26°)	−11° (12°)	−19° (−3°)	−27° (−17°)	−35° (−31°)	−43° (−45°)	−51° (−60°)	−59° (−74°)	−67° (−88°)

Wind speed, kmph (mph)

Frostbite times: 30 min.   10 min.   5 min.

Figure 5.3
Wind chill chart for various temperatures and wind velocities

1. Use the wind-chill chart in Figure 5.3 to determine the wind-chill temperature for each of the following examples:

    °C   (°F)

    a) Wind speed: 24 kmph, air temperature: −34°C = wind-chill temp: _____

    b) Wind speed: 48 kmph, air temperature: −7°C = wind-chill temp: _____

    c) Wind speed: 8 kmph, air temperature: +4°C = wind-chill temp: _____

    d) Wind speed: 56 kmph, air temperature: −23°C = wind-chill temp: _____

2. Competitive downhill ski racers are subjected to severe wind chill, as are average skiers and snowboarders to a lesser degree. Assuming a downhill racer is going 80 kmph (50 mph), which is coasting on some runs, and

    the air temperature is −18°C (0°F), what is the wind chill they are feeling on any exposed skin? _____

    What is their time to experience frostbite given these conditions? _____. In televised coverage,

    what kind of outfits do these racers wear for protection? _____

    _____

    _____

    _____

3. Use the heat index chart in Figure 5.4 to determine the heat index temperature for each of the following examples: °C (°F)

   a) Air temperature: 37.8°C, relative humidity 5% = heat index temp: _____

   b) Air temperature: 32.2°C, relative humidity 80% = heat index temp: _____

   c) Air temperature: 32.2°C, relative humidity 90% = heat index temp: _____

   d) Air temperature: 43.3°C, relative humidity 10% = heat index temp: _____

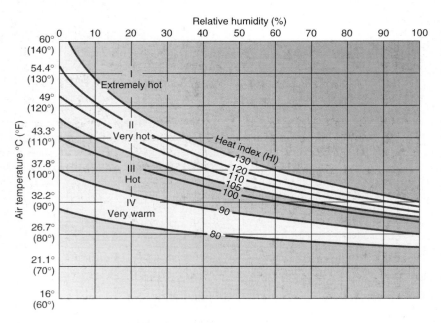

Level of concern	Category	Heat Index Apparent Temperature	General Effect of Heat Index on People in High-Risk Groups
Extreme danger	I	54°C (130°F) or higher	Heat/sunstroke highly likely with continued exposure
Danger	II	41° – 54°C (105° – 130°F)	Sunstroke, heat cramps, or heat exhaustion likely and heatstroke possible with prolonged exposure and/or physical activity
Extreme caution	III	32° – 41°C (90° – 105°F)	Sunstroke, heat cramps, and heat exhaustion possible with prolonged exposure and/or physical activity
Caution	IV	27° – 32°C (80° – 90°F)	Fatigue possible with prolonged exposure and/or physical activity

Figure 5.4
Heat index for various temperatures and relative humidity levels.

4. List the heat index categories would you experience if the temperature stayed a constant 35° C (95° F), but

   the relative humidity dropped from 90% down to 10%. _____

   _____

Name: _____   Laboratory Section:_____

Date: _____   Score/Grade: _____

Lab Exercise 6

Temperature Patterns

Principal controls and influences upon temperature patterns include latitude, elevation, cloud cover, land-water heating differences, ocean currents and sea-surface temperatures, and general surface conditions. A variety of temperature regimes worldwide affect cultures, decision making, and resources con- sumption. Global temperature patterns apparently are changing in a warming trend and are the subject of much scientific, geographic, and political interest. In Lab Exercise 6, we analyze the spatial patterns of global temperatures. Lab Exercise 6 features four sections.

Key Terms and Concepts

average annual temperature
continentality
environmental lapse rate
isotherm
land-water heating differences

marine (maritime)
normal lapse rate
specific heat
temperature range

Objectives

After completion of this lab, you should be able to:

1. *Contrast* and *compare* the temperature patterns of La Paz and Concepción, Bolivia, and *determine* the effects of elevation on temperatures.
2. *Compare* and *contrast* the temperature patterns of Vancouver, British Columbia, and Winnipeg, Mani- toba, and *determine* the differing effects of marine and continental factors.
3. *Plot* and *graph* temperature data on climographs for selected stations.
4. *Construct* isotherms on a U.S.-Canadian map and *analyze* resultant temperature patterns portrayed.
5. *Analyze* global temperature patterns by plotting temperature profiles.

Materials/Sources Needed

pencil
color pencils
highlighter pen
calculator

Lab Exercise and Activities

✻ SECTION 1

Elevation and Temperature

Within the troposphere, temperatures decrease with increasing altitude above Earth's surface: the **normal lapse rate** of temperature change with altitude is 6.4 C°/1000 m or 3.5 F°/1000 ft. Worldwide, mountainous areas experience lower temperatures than do regions nearer sea level, even at similar latitudes. The consequences are that average air temperatures at higher elevations are lower, nighttime cooling increases, and the temperature range between day and night and between areas of sunlight and shadow also increases. Temperatures may decrease noticeably in the shadows and shortly after sunset. Surfaces both gain heat rapidly and lose heat rapidly to the thinner atmosphere.

As we observed in Lab Exercise 5, the distribution of intercepted solar energy exhibits an overall imbalance among equatorial, midlatitude, and polar locations. The equatorial regions exceed the polar regions by 2.5 times in terms of insolation received. As an interesting example of the effects of elevation, let's examine two cities in Bolivia, La Paz and Concepción.

Temperature data are presented in Table 6.1 for La Paz and Concepción. Both stations are approximately the same distance south of the equator but differ in elevation. La Paz is at 4103 m (13,461 ft), whereas Concepción is at 490 m (1608 ft) above sea level. The hot, humid climate of Concepción at its much lower elevation stands in marked contrast to the cool, dry climate of highland La Paz.

People living around high-elevation La Paz actually grow wheat, barley, and potatoes—crops characteristically grown in the cooler midlatitudes at lower elevations. These crops do well despite the fact that La Paz is 4103 m above sea level. The combination of *elevation* (moderating temperatures) and *equatorial location* (producing higher Sun altitude and consistent daylength) guarantee La Paz these temperature conditions, averaging about 9°C (48°F) for every month. Such moderate temperature and moisture conditions lead to the formation of more fertile soils than those found in the warmer, wetter climate of Concepción.

---

1. Using the temperature graphs provided in Figure 6.1, plot the data from Table 6.1 for these two cities. Use a smooth curved *line graph* to portray the temperature data. Calculate the ***mean* (average)** *annual temperature* and the ***temperature range*** (difference between the highest and lowest) for each city.

2. Why are the temperatures at La Paz more moderate in every month and so consistent overall as compared to

   Concepción? _____

   _____

   _____

   _____

3. Recall that the *normal lapse rate* of temperature change with altitude is 6.4 C°/1000 m or 3.5 F°/1000 ft. Calculate the difference in elevation between La Paz and Concepción, and calculate what the difference in their *mean (average) annual temperatures* should be based on cooling at the normal lapse rate. Is the actual difference between the mean annual temperatures for these two cities higher, lower, or the same as that

   produced by calculating average normal lapse rate conditions? (Show your work.) _____

   _____

   _____

4. The annual march of the seasons and the passage of the subsolar point between the tropics of Cancer and Capricorn affect these stations. Can you detect from your temperature graphs these seasonal effects? Explain.

_____

_____

La Paz, Bolivia

Avg. Annual Temperature (°C, °F): _____

Annual Temp. Range (°C, °F): _____

Concepción, Bolivia

Avg. Annual Temperature (°C, °F): _____

Annual Temp. Range (°C, °F): _____

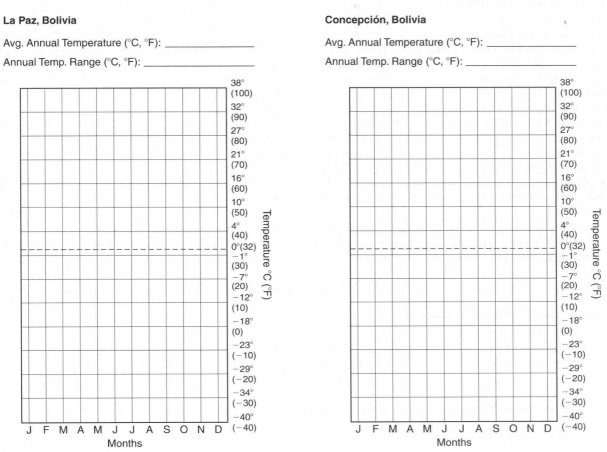

Figure 6.1
Temperature graphs for La Paz and Concepción, Bolivia

Table 6.1
Temperature data for La Paz and Concepción, Bolivia
La Paz, Bolivia: pop. 877,363, lat. 16.5° S, long. 68.17° W, elev. 4103 m (13,461 ft).

	Jan	Feb	Mar	Apr	May	Jun	Jul	Aug	Sep	Oct	Nov	Dec
Avg. Temp.°C	9.0	9.0	9.0	9.0	8.0	7.0	8.0	8.0	9.0	10.0	10.0	10.0
(°F)	(48.2)	(48.2)	(48.2)	(48.2)	(46.4)	(44.6)	(46.4)	(46.4)	(48.2)	(50.0)	(50.0)	(50.0)

Concepción, Bolivia: pop. 8,221, lat. 16.25° S, long. 62.05° W, elev. 490 m (1608 ft).

	Jan	Feb	Mar	Apr	May	Jun	Jul	Aug	Sep	Oct	Nov	Dec
Avg. Temp.°C	25.0	26.0	25.0	24.0	23.0	19.0	22.0	24.0	27.0	27.0	26.0	26.0
(°F)	(77.0)	(78.8)	(77.0)	(75.2)	(73.4)	(66.2)	(71.6)	(75.2)	(80.6)	(80.6)	(78.8)	(78.8)

Marine versus Continental Effects

The irregular arrangement of landmasses and water bodies on Earth contributes to the overall pattern of temperature. The physical nature of the substances themselves—rock and soil versus water—is the reason for these **land-water heating differences**. More moderate temperature patterns are associated with water bodies compared to more extreme temperatures inland.

These contrasts in temperatures are the result of the land-water temperature controls: evaporation, transmissibility, **specific heat**, movement, ocean currents, and sea-surface temperatures. The term **marine**, or *maritime*, is used to describe locations that exhibit the moderating influences of the ocean, usually along coastlines or on islands. **Continentality** refers to the condition of areas that are less affected by the sea and therefore have a greater range between maximum and minimum temperatures diurnally (daily) and yearly.

The Canadian cities of Vancouver, British Columbia, and Winnipeg, Manitoba, exemplify these marine and continental conditions. Both cities are at approximately 49° N latitude. Respectively, they are at sea level and 248 m (814 ft) elevation. However, Vancouver has a more moderate pattern of average maximum and minimum temperatures than Winnipeg. Vancouver's annual range of 11.1 C° (20.0 F°) is far below the 38.8 C° (70.0 F°) temperature range in Winnipeg. In fact, Winnipeg's continental temperature pattern is more extreme in every aspect than that of maritime Vancouver.

1. Using the data given in Table 6.2, plot the temperatures for these two cities and portray with a smooth curved *line graph* on the temperature graph in Figure 6.2. Calculate the average annual temperature and temperature range for each city.

2. Compare and contrast the marine temperature regime of Vancouver with the continental regime of Winnipeg.

   What significant differences do you note? _____

   _____

   _____

3. How many months register average temperatures below freezing in Winnipeg? _____

   How many months average below freezing in Vancouver?_____

   Explain. _____

   _____

Vancouver, British Columbia

Avg. Annual Temperature (°C, °F): _____

Annual Temp. Range (°C, °F): _____

Winnipeg, Manitoba

Avg. Annual Temperature (°C, °F): _____

Annual Temp. Range (°C, °F): _____

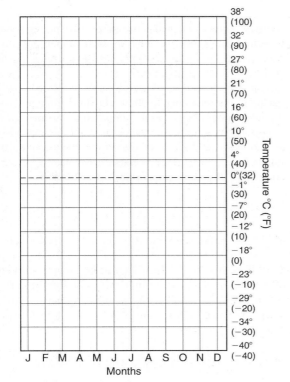

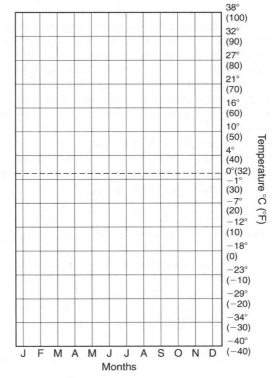

Figure 6.2
Temperature graphs for Vancouver, British Columbia, and Winnipeg, Manitoba, Canada

Table 6.2
Temperature data for Vancouver, British Columbia, and Winnipeg, Manitoba, Canada
Vancouver, British Columbia: pop. 578,041, lat. 49.18° N, long. 123.16° W, elev. sea level.

	Jan	Feb	Mar	Apr	May	Jun	Jul	Aug	Sep	Oct	Nov	Dec
Avg. Temp.°C	3.3	4.4	5.6	7.8	10.0	12.2	13.9	14.4	12.2	9.4	6.7	4.4
(°F)	(37.9)	(39.9)	(42.1)	(46.0)	(50.0)	(54.0)	(57.0)	(57.9)	(54.0)	(48.9)	(44.1)	(39.9)

Winnipeg, Manitoba: pop. 633,451 lat. 49.9° N. long. 97.23° W, elev. 248 m (813.6 ft).

	Jan	Feb	Mar	Apr	May	Jun	Jul	Aug	Sep	Oct	Nov	Dec
Avg. Temp.°C	−19.4	−16.7	−8.9	3.3	11.1	16.7	19.4	17.8	12.2	5.0	−5.6	−14.4
(°F)	(−2.9)	(1.9)	(16.0)	(37.9)	(52.0)	(62.1)	(66.9)	(64.0)	(54.0)	(41.0)	(21.9)	(6.1)

Working with Isotherms and Temperature Maps

Both the pattern and variety of temperature on Earth are outputs of the global energy system. Temperature maps portray the interaction of the principal temperature controls: latitude, altitude, cloud cover, and land-water heating differences. To depict the spatial aspect of data, geographers utilize *isopleths* (also called *isolines*)— lines drawn on a map connecting points of equal value, such as contour lines on a topographic map that show points of equal elevation (Lab Exercise 14).

An isopleth on temperature maps is known as an *isotherm*; pressure maps = *isobars*; precipitation maps = isohyets; equal-wind velocity = *isotachs* . An isotherm

connects points of equal temperature and presents the pattern of temperature for analysis. Geographers are concerned with the spatial distribution of temperatures, and isotherms facilitate this analysis.

In Lab Exercise 14, ❋ SECTION 2, you will construct contour lines to show elevation change. Isotherms, connecting equal temperature values, are drawn in much the same way. Construct the appropriate isotherms at 5 C° intervals on the weather map in Figure 6.3. The surface weather map data were recorded at 7:00 A.M. E.S.T., on October 30, 1993.

---

1. The 0°C (32°F) isotherm is already drawn for you (highlight it to make it distinctive). You may find it helpful to use color pencils to "group" the temperature intervals (5 C° intervals, above and below 0°C). Use a pencil to lightly sketch the isotherms, then use a pen to darken the lines, which should be smooth curved lines. (See your geography text or the weather page of the newspaper for examples of isothermal maps.) Begin with the −5°C isotherm that is drawn for you, then draw the −10°C and −15°C isotherm. Now, continue by completing the +5°C, +10°C, +15°C (drawn for you), +20°C, and finally, the +25°C isotherm (also drawn for you) across central Florida.

2. What kind of temperature pattern can be observed over the Great Plains (northern Texas, southeastern Colorado to the Dakotas). What do you think is producing this condition? _____

    _____

3. Describe and explain the temperature gradient over Nevada and Utah. (How rapidly are the temperatures changing from one place to another?) _____

    _____

4. Describe the temperature gradient over most of Texas. _____

    _____

5. Describe and explain the temperature gradient over the southeastern states. _____

    _____

6. What is the temperature range on this map? Calculate the difference between the maximum (highest) and minimum (lowest) temperature reading. In the spaces below, fill in the readings; use a map or atlas to determine the city (station) where the data was collected.

    Maximum temperature: _____ Station (City) _____

    Minimum temperature: _____ Station (City) _____

    Range (express in C° and F°): _____

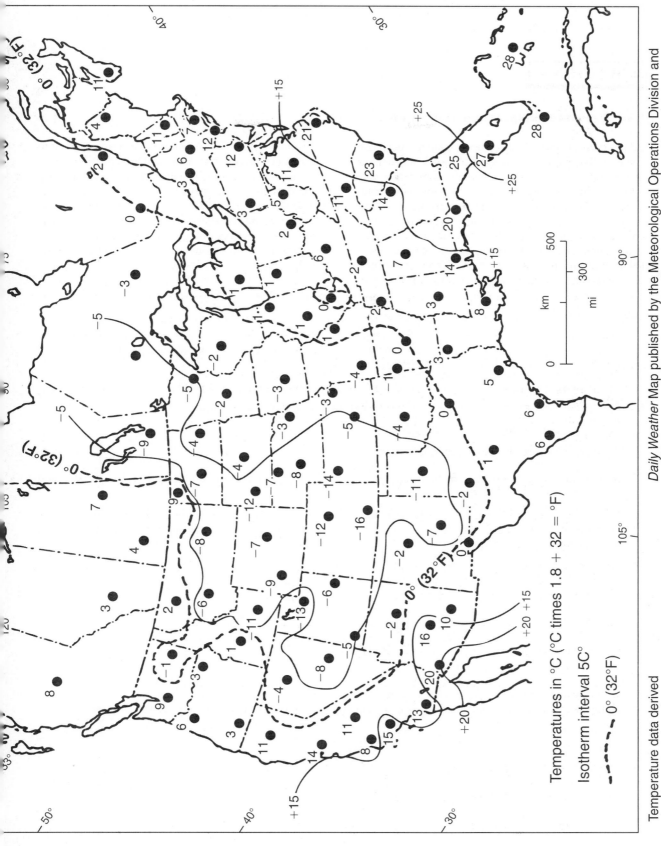

Figure 6.3
Isotherm map

Daily Weather Map published by the Meteorological Operations Division and Climate Analysis, National Meteorological Center, National Weather Service, National Oceanic and Atmospheric Administration, Department of Commerce, 1993.

Temperatures in °C (°C times 1.8 + 32 = °F)

Isotherm interval 5C°

— — 0° (32°F)

Temperature data derived from the *Daily Weather Map* for October 30, 1993, 7:00 A.M. E.S.T.

Global Surface Temperature Map Analysis

Just as closer contour lines on a topographic map (Lab Exercise 14) mark a steeper slope, so do closer isotherms denote steepness in the temperature gradient (change across Earth's surface). In Figure 6.4a (January), note the spacing of the isotherms along the 90° W meridian between 40°–70° N latitudes (steep temperature gradient) and between 40°–70° S latitudes (gradual temperature gradient). Isotherms spaced at greater distances from one another mark a more gradual temperature change. Relate these patterns to the distribution of land and water surfaces—water is generally more moderate, land more extreme, in temperature distributions.

As you use the maps in Figures 6.4a and 6.4b, remember that these maps use a 3°C interval (example: 0°C, 3°C, 6°C, 9°C, etc.). If the data points you are plotting on the charts in Figures 6.5a, 6.5b, 6.6a, and 6.6b fall between two isotherms, you can interpolate (estimate) the temperature value to mark on the graph.

1. Temperature profiles, like topographic profiles (Lab Exercise 13), illustrate how rapidly temperatures change over Earth's surface. Use the four graphs in Figures 6.5a, 6.5b, 6.6a, and 6.6b to plot temperature data along two parallels and two meridians as noted; then complete the temperature profiles with a line graph connecting the plotted data points:

   a) January (Figure 6.4a), along 50° N parallel (already plotted and drawn for you)

   b) January (Figure 6.4a), along 90° W meridian (already plotted and drawn for you)

   c) July (Figure 6.4b), along 40° N parallel

   d) July (Figure 6.4b), along 60° E meridian

   After completing the two graphs in Figures 6.6a and 6.6b, label that part of your plots that are over land and over water, then complete the following.

2. Using these maps and graphs, describe and explain average temperatures and temperature gradients: Where are temperatures higher/lower? Where are gradients steep or gentle? Contrast temperatures over landmasses and oceans at the same latitudes.

   a) January: _____

   _____

   b) July: _____

   _____

3. Describe and explain the differences in temperature patterns between the Northern and Southern Hemispheres in January and July. What is the general pattern of isotherms? Where are the extremes? Where is it more moderate? _____

   _____

   _____

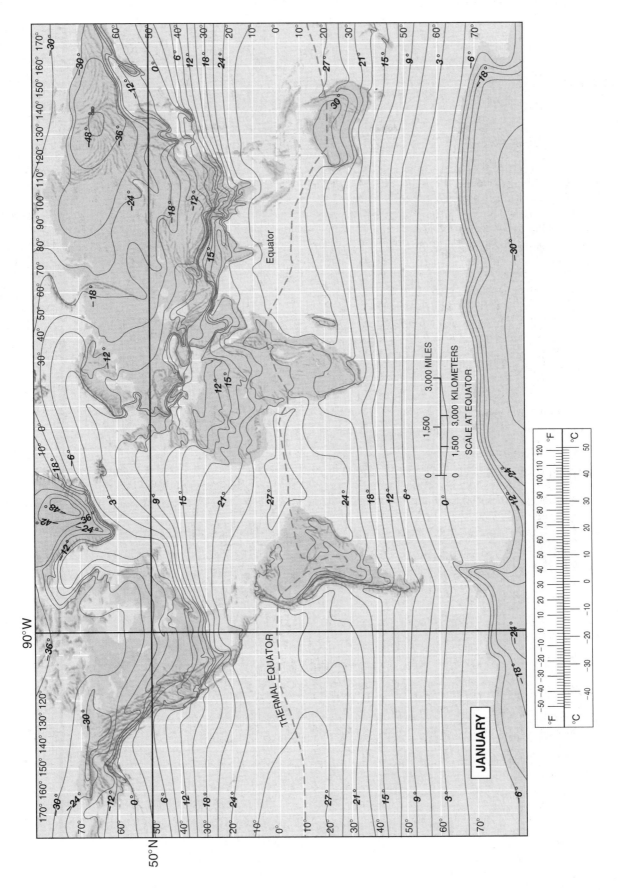

Figure 6.4a
(a) Global temperatures for January

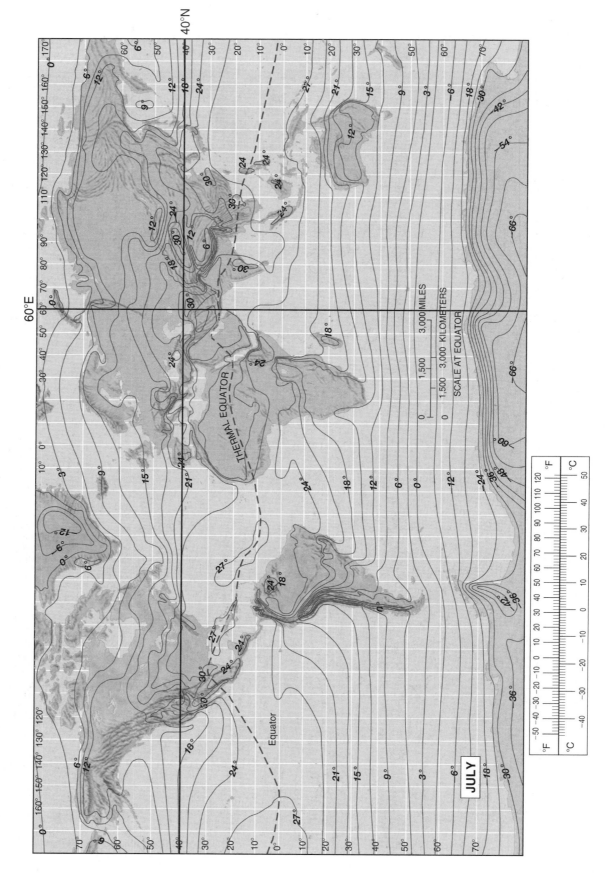

Figure 6.4b
(b) Global temperatures for July

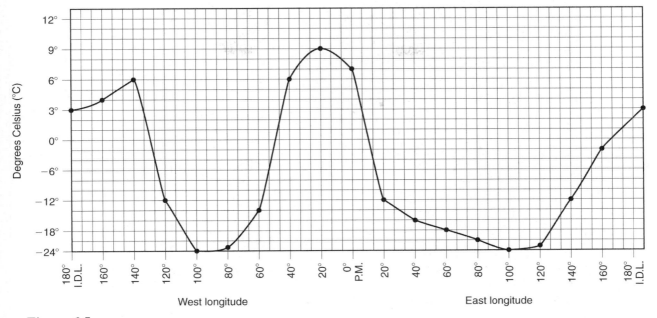

Figure 6.5a
(a) Plot of average temperatures along 50° N latitude for January

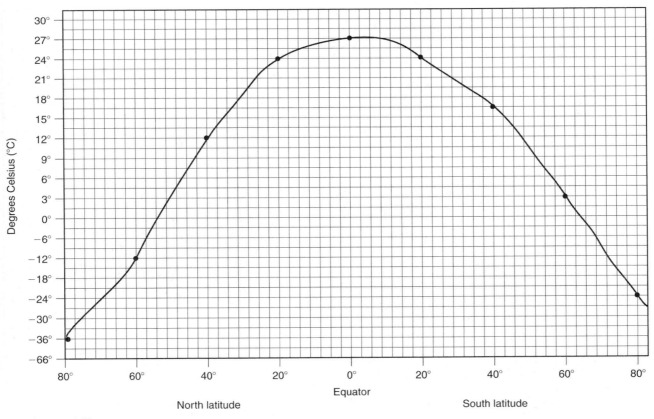

Figure 6.5b
(b) Plot of average temperatures along 90° W longitude for January

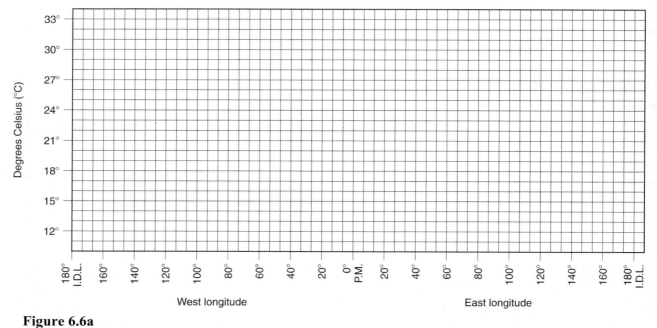

Figure 6.6a
(a) Plot of average temperatures along 40° N latitude for July

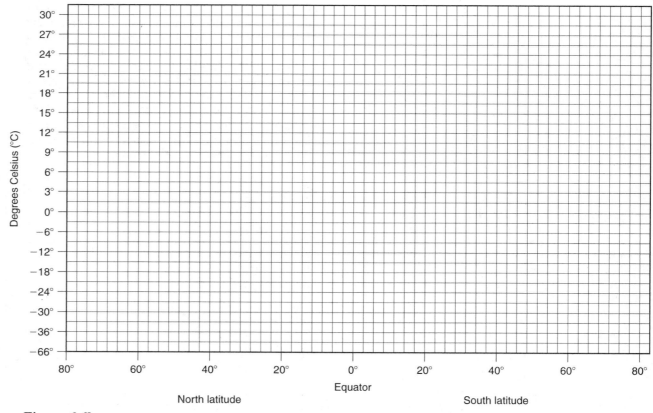

Figure 6.6b
(b) Plot of average temperatures along 60° E longitude for July

Name: _____   Laboratory Section:_____

Date: _____   Score/Grade: _____

Lab Exercise 7

Earth's Atmosphere: Pressure Profiles and Pressure Patterns

The gases that make up air create pressure through their motion, size, and number. This pressure is exerted on all surfaces in contact with the air. The weight of the atmosphere, or **air pressure**, exerts an average force of approximately 1 kg/cm^2 (14.7 lb/in^2) at sea level. Under the acceleration of gravity, air is compressed and therefore denser near Earth's surface. The atmosphere rapidly thins with increased altitude, a decrease that is measurable because air exerts its weight as pressure. Consequently, over half the total mass of the atmosphere is compressed below 5500 m (18,000 ft), 75% is compressed below 10,700 m (35,105 ft), and 90% is below 16,000 m (52,500 ft). All but 0.1% is accounted for at an altitude of 50,000 m (163,680 ft) or 50 km (31 mi).

Any instrument that measures air pressure is called a **barometer**. One type of barometer uses a column of mercury that is counterbalanced by the mass of surrounding air exerting an equivalent pressure on a vessel of mercury to which the column is attached. Evangelista Torricelli developed the **mercury barometer**. Another type of barometer, the **aneroid barometer**, is a small chamber that is partially emptied of air, sealed, and connected to a mechanism that is sensitive to changes in air pressure. As air pressure varies, the mechanism responds.

Altimetry is the measurement of altitude using air pressure. A *pressure altimeter* is an instrument that measures altitude based on a strict relationship between air pressure and altitude. Air pressure is measured within the altimeter by an aneroid barometer capsule with the instrument graduated in increments of altitude. A pilot must constantly adjust the altimeter, or "zero" the altimeter, during flight as atmospheric densities change with air temperature. Another type of altimeter sends and receives radio wavelengths between the plane and the ground to determine altitude—a *radio altimeter*. In this lab we examine Earth's atmospheric pressure patterns. Lab Exercise 7 features three sections.

Key Terms and Concepts

air pressure
altimetry
aneroid barometer
barometer

isobars
mercury barometer
pressure gradient
standard atmosphere

Objectives

After completion of this lab, you should be able to:

1. *Demonstrate* pressure trends with altitude.
2. *Explain* the standard atmosphere concept and *plot* key elements on a graph.
3. *Locate* a properly installed barometer (and, if necessary, a thermometer) and *obtain* data over a three-day period.
4. *Interpret* global pressure patterns and *construct* zonal and meridional pressure profiles.
5. *Analyze* global pressure patterns by plotting pressure profiles.

Materials/Sources Needed

pencil
color pencils
calculator
textbook

Lab Exercise and Activities

✳ SECTION 1

Air Pressure

Normal sea level pressure is expressed as 1013.2 mb (millibars) of mercury (a way of expressing force per square meter of surface area). At sea level standard atmospheric pressure is expressed in several ways:

- 14.7 lb/in^2
- 29.9213″ of Hg (mercury)
- 1013.250 millibars
- 101.325 kilopascals (kilopascal = 10 millibars)

Some convenient conversions are helpful:

- 1.0 in. (Hg) = 33.87 mb = 25.40 mm (Hg) = 0.49 lb/in^2
- 1.0 mb = 0.0295 in. (Hg) = 0.75 mm (Hg) = 0.0145 lb/in^2

The **standard atmosphere** for pressure in millibars and altitude in kilometers is given in Table 7.1. This is used in the activity that follows the table.

Table 7.1
Standard atmosphere for pressure and altitude

Altitude (km)	Pressure (mb)	Altitude (km)	Pressure (mb)
0.00	1013.25	10.00	264.99
0.50	954.61	12.00	193.99
1.00	898.76	14.00	141.70
1.50	845.59	16.00	103.52
2.00	795.01	18.00	75.65
2.50	746.91	20.00	55.29
3.00	701.21	25.00	25.49
4.00	616.60	30.00	11.97
5.00	540.48	35.00	5.75
6.00	472.17	40.00	2.87
7.00	411.05	50.00	0.79
8.00	356.51	60.00	0.23
9.00	308.00	70.00	0.06

1. Using the graph in Figure 7.1, *plot* the standard atmosphere of air pressure decrease with altitude presented in Table 7.1. After completing the plot, connect the data points with a line to complete the pressure profile of the atmosphere. The data points from 0 to 5 km are plotted for you.

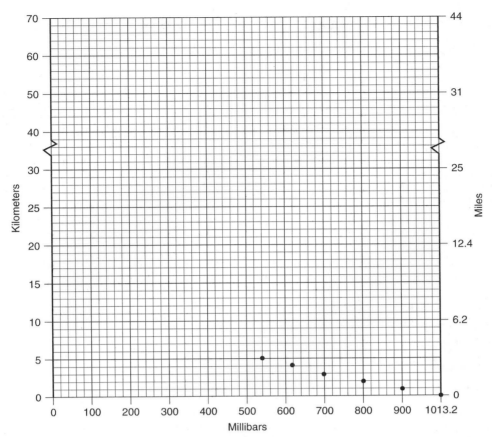

Figure 7.1
Atmospheric pressure profile graph—the standard atmosphere—from the surface to 70 km.

2. The information in Table 7.1 allows a determination of the rate of pressure decrease with altitude, which is not at a constant rate. Remember that half of the weight of the total atmosphere occurs below 5500 m (18,000 ft); at that altitude only about half of the total atoms and molecules of atmospheric gases remain to form the mass of the atmosphere. Determine the decrease in pressure between the following altitudes. Express the difference in millibars and inches of mercury. (Conversions are presented earlier in this section.)

1 km interval difference in pressure:

0 and 1 km ___[114.49]___ mb; ___[3.38]___ in.

2 and 3 km _____ mb; _____ in.

5 and 6 km _____ mb; _____ in.

8 and 9 km _____ mb; _____ in.

9 and 10 km _____ mb; _____ in.

10 km interval difference in pressure:

0 and 10 km ___[748.26]___ mb; ___[22.07]___ in.

10 and 20 km _____ mb; _____ in.

20 and 30 km _____ mb; _____ in.

40 and 50 km _____ mb; _____ in.

60 and 70 km _____ mb; _____ in.

Show conversion work here: _____ [114.49 mb×.0295 in/mb = 3.38 in] _____

_____

3. Using the graph you prepared in Figure 7.1, <u>approximate</u> the answers to the following (assuming standard atmosphere conditions):
   a) Mount Everest's summit is 8850 m (29,035 ft) above sea level. What is the barometric pressure there according to the standard atmosphere? _____ [320 mb] _____

b) Mount McKinley, 6194 m (20,320 ft); air pressure at the summit? _____

c) Mount Whitney, 4418 m (14,494 ft); air pressure at the summit? _____

d) Yellowstone Lake, Yellowstone N.P., 2356 m (7731 ft); air pressure? _____

e) The Petronas Towers I and II, Kuala Lumpur, Malaysia, 452 m (1483 ft); air pressure? _____

f) In a commercial airliner taking you from San Francisco to New York at 12,000 m (39,400 ft), what

percentage of atmospheric pressure is below your plane? _____

What percentage of atmospheric pressure resides above your flight altitude? _____

4. Why does atmospheric pressure decrease so rapidly with altitude? _____

_____

_____

_____

✳ SECTION 2

Pressure Readings

Lab Exercise 5, ✳ SECTION 3 asked you to record air temperature (and also air pressure) for three days. If you did not record barometric readings at that time, locate a barometer either at the college or university you are attending, or perhaps at home. Many students find that someone in the family was given an aneroid barometer years ago; perhaps one that no one quite knows how to use. If you do not have access to a barometer, find a source for information about barometric pressure (weather broadcasts, Internet sources). At the same time you may want to record air temperature. As before, the goal is to locate and use a reliable source of barometric pressure. This will be an asset in your life, travel, and activities.

1. You started collecting data on the Weather Calendar in the Prologue Lab. As you learn more about elements of weather, you should begin to apply what you learn to the weather that you experience on a daily basis. Collecting and working with weather data will enhance your awareness and understanding of weather processes.

For at least three days, record the air pressure (and air temperature) at approximately the same time of day, if possible. See if you can detect a trend in air pressure changes or a relationship between air pressure and other atmospheric processes. *Record* your observations below.

Date and Pressure	Place of observation	Time of day	Barometer used
Day 1:	Place:	Time:	Type:
Day 2:	Place:	Time:	Type:
Day 3:	Place:	Time:	Type:

2. Weather conditions and state of the sky at the time of your observation:

Day 1: _____

_____

Day 2: _____

_____

Day 3: _____

_____

❋ SECTION 3

Air Pressure Map Analysis

High- and low-pressure areas that exist in the atmosphere principally result from unequal heating at Earth's surface and from certain dynamic forces in the atmosphere. A map of pressure patterns is accomplished by using **isobars**, lines that connect points of equal pressure. As with other isolines that we have used—contour lines and isotherms—the distance between isobars indicates the degree of pressure difference, or **pressure gradient**, which is the change in atmospheric pressure over distance between areas of higher pressure and lower pressure. Isobars facilitate the spatial analysis of pressure patterns and are key to weather map preparation, interpretation, and forecasting.

Sea-level atmospheric pressure averages for January and July are portrayed in Figures 7.2 and 7.3. These maps use long-term measurements from surface stations and ship reports generally taken from the 1950s to 1970s, with some ship data going back to the last half of the 19th century.

Surface air pressure map analysis:

Just as closer contour lines on a topographic map mark a steeper slope and closer isotherms (Lab Exercise 6) mark steeper temperature gradients, so do closer isobars denote steepness in the pressure gradient. Isobars spaced at greater distances from one another mark a more gradual pressure gradient, one that creates a slower air flow. A steep gradient causes faster air movement from a high-pressure area to a low-pressure area.

In Figure 7.2 (January), note the spacing of the isobars along the 90° W meridian between 40°–70° N latitudes (gradual pressure gradient—weaker winds) and between 40°–70° S latitudes (steeper pressure gradient—stronger winds). As you use the maps in Figures 7.2 and 7.3, remember these maps use a 2-mb interval (e.g., 1020 mb, 1018 mb, 1016 mb, etc.). If the data points you are plotting fall between two isobars, you can interpolate (estimate) the pressure value to mark on the graph.

1. Use the four graphs in Figures 7.4a, 7.4b, 7.5a, and 7.5b to plot pressure data along two parallels and two meridians as noted; then complete the pressure profile with a line graph connecting the plotted data points (for every 20°):

    a) January (Figure 7.2), along 50° N parallel (already done for you)

    b) January (Figure 7.2), along 90° W meridian

    c) July (Figure 7.3), along 40° N parallel

    d) July (Figure 7.3), along 60° E meridian

After completing the plotting of pressure data on graphs in Figures 7.4a, 7.4b, 7.5a, and 7.5b, complete the following.

2. Using these maps and your textbook explain the patterns of average atmospheric pressure over the landmasses. Are the patterns of high and low pressure caused by thermal or dynamic forces?

    a) January: _____

    _____

    _____

    b) July: _____

    _____

    _____

3. Using Figures 7.2 and 7.3, label the primary subtropical high-pressure and low-pressure cells in the northern hemisphere. Draw and label the primary wind systems in the mid-latitudes and tropics.

For Figure 7.3, draw arrows indicating the monsoonal flow of air from the Indian Ocean northward into India and the Tibetan Plateau.

4. In south and central Asia, given the pressure patterns you observe on the maps, what wind patterns (direction, velocity) would you expect to find and <u>why</u>?

   a) January: _____

_____

_____

   b) July: _____

_____

_____

5. During the Southern Hemisphere summer, describe the pressure gradient over the Southern Ocean (South Pacific and South Atlantic Oceans). What do you think this gradient produces? Discuss with others in your lab to find if there is a popular term for these latitudes of such a pressure gradient.

_____

_____

_____

6. Refer back to the global air temperature profiles you plotted in ✳ SECTION 4 of Lab Exercise 6 (temperature maps in Figure 6.4a January and 6.4b July, and your graphs in Figure 6.5a, b and Figure 6.6a, b). Those temperature profiles in Lab Exercise 6 and these pressure profiles you just completed for this exercise are along the same parallels and meridians.

Comparing the two sets of graphs (Figures 6.5 and 6.6 with Figures 7.4 and 7.5), what correlation can be seen between global temperature and pressure patterns? Select a few areas that seem to you to illustrate a link between your graphs.

_____

_____

_____

_____

_____

_____

_____

_____

The following pages contain:

        Figure 7.2: January world pressure map
        Figure 7.3: July world pressure map
        Figure 7.4a, b: plots for 50° N and for 90° W (7.4a and 7.4b are completed for you)
        Figure 7.5a, b: plots for 40° N and 60° E

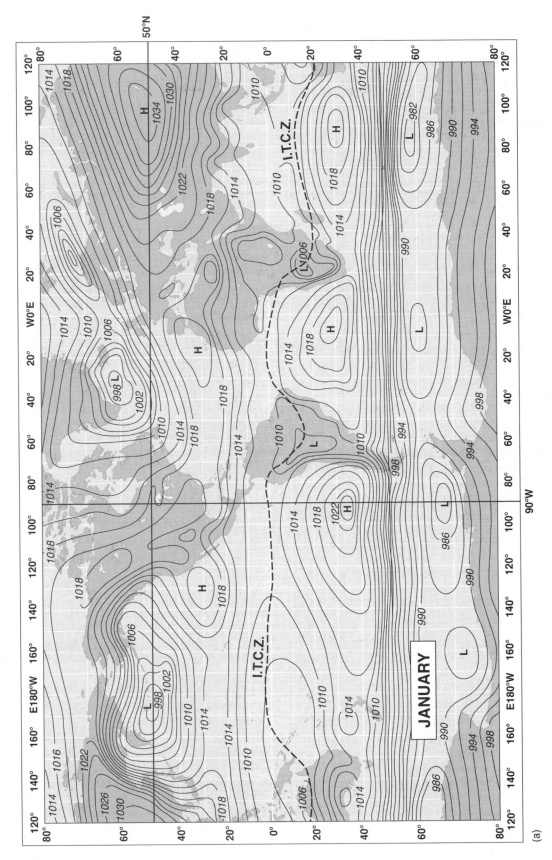

Figure 7.2
Global barometric pressure for January

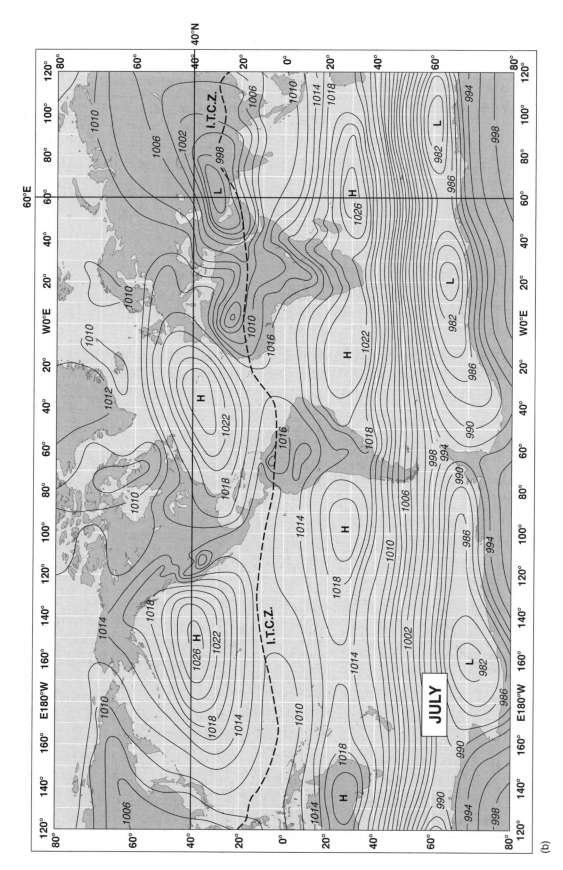

Figure 7.3
Global barometric pressure for July

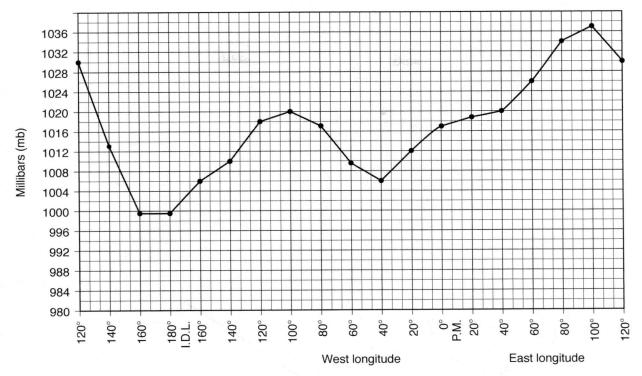

Figure 7.4a
(a) Plot of average air pressure along 50° N latitude for January.

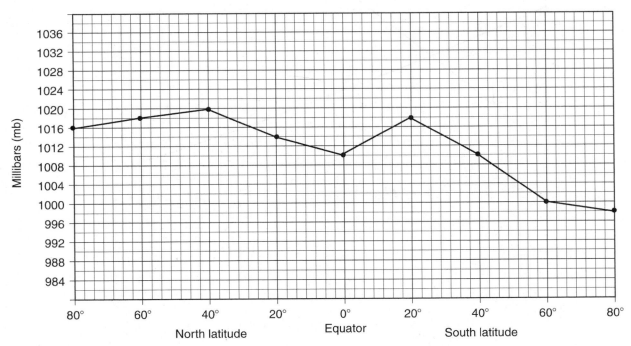

Figure 7.4b
(b) Plot of average air pressure along 90° W longitude for January.

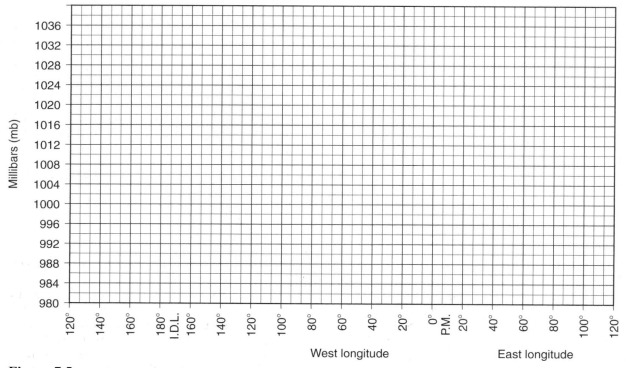

Figure 7.5a
(a) Plot of average air pressure along 40° N latitude for July.

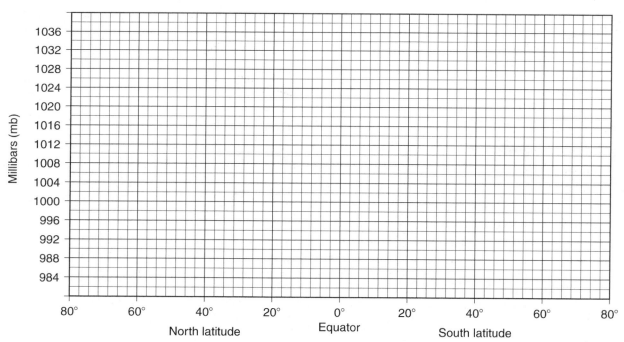

Figure 7.5b
(b) Plot of average air pressure along 60° E longitude for July.

Name: _____   Laboratory Section: _____

Date: _____   Score/Grade: _____

Lab Exercise 8

Atmospheric Humidity, Stability, and Adiabatic Processes

The powerful weather drama we witness daily in the sky is fueled by insolation as heat energy is delivered to the atmosphere by water vapor. The water vapor content of air is termed humidity. The capacity of air to hold water vapor is primarily a function of temperature, both the temperature of the air and of the water vapor, which are usually the same. Warmer air has a greater capacity for holding water vapor, whereas cooler air has a lesser capacity. To determine the energy available for weather activities, it is essential to know the water vapor content

of air. Humidity is easily measurable with instruments that we use in this exercise.

Humidity and temperature characteristics of an air mass are key parameters in determining the stability of the air. Conditions in the environment determine whether a mass of air will lift, cool by expansion, and become saturated—unstable conditions—or whether it will resist displacement—stable conditions. This exercise examines atmospheric humidity, stability aspects of air masses, and adiabatic processes. Lab Exercise 8 has six sections.

Key Terms and Concepts

adiabatic
dew-point temperature
dry adiabatic rate (DAR)
environmental lapse rate
hair hygrometer
humidity
lifting condensation level
moist adiabatic rate (MAR)
normal lapse rate

orographic lifting
rain shadow
relative humidity
saturated
sling psychrometer
specific humidity
stability
vapor pressure

Objectives

After completion of this lab, you should be able to:

1. *Define* humidity and *describe* the several ways of expressing measurements of humidity.
2. *Demonstrate* the use of a sling psychrometer to *calculate* and *determine* relative humidity.
3. *Contrast* environmental lapse rates with adiabatic processes to *calculate* the stability of an air mass.
4. *Define* dew-point temperature, saturation, and lifting condensation level.
5. *Identify* the processes that lead to condensation, cloud development, and precipitation.
6. *Evaluate* orographic lifting effects and apply adiabatic heating and cooling rates.

Materials/Sources Needed

pencils
calculator
color pencils

hair hygrometer (if available for demonstration)
psychrometric tables (provided)
sling psychrometer

Lab Exercise and Activities

❊ SECTION 1

Relative Humidity

Next to temperature and pressure, the most common piece of information in local weather broadcasts is **relative humidity**. Relative humidity is not a direct measurement of water vapor. Rather, it is a ratio (expressed as a percentage) of the amount of water vapor that is actually in the air (*content*), compared to the maximum water vapor the air could hold at a given temperature (*capacity*). If the air is relatively dry in comparison to its capacity, the percentage is lower; if the air is relatively moist, the percentage is higher.

Relative humidity is calculated as follows:

$$\text{Relative humidity} = \frac{\text{actual water vapor } \textit{content} \text{ of the air}}{\text{maximum water vapor } \textit{capacity} \text{ of the air}} \times 100$$

EXAMPLE:

An air mass at 15°C has a water vapor pressure content of 10.2 mb. The saturation vapor pressure (maximum capacity) of the air to hold water vapor at that temperature is 17.0 mb. Therefore, the relative humidity is:

$$\text{Relative humidity} = \frac{10.2 \text{ mb}}{17.0 \text{ mb}} = .60 \times 100 = \textbf{60\%}$$

Relative humidity varies due to evaporation, condensation, or temperature changes, all of which affect both the *content* and the *capacity* of the air to hold water vapor. Air is said to be **saturated**, or full, if it is holding all the water vapor that it can hold at a given temperature; under such conditions, the net transfer of water molecules between any surface and the air achieves a *saturation equilibrium*. Saturated air has a relative humidity of 100%. Saturation indicates that any further addition of water vapor (change in content) or any decrease in temperature (change in capacity) will result in active condensation.

The temperature at which a given mass of air becomes saturated is termed the **dew-point temperature**. In other words, *air is saturated when the dew-point temperature and the air temperature are the same*. There are several ways to express humidity, relative humidity, and saturation concepts. Here we examine **vapor pressure** as a way of expressing humidity; in ❊ SECTION 5 we use **specific humidity** as another expression of humidity.

That portion of total air pressure that is made up of water vapor molecules is termed **vapor pressure** and is expressed in millibars (mb). In terms of vapor pressure the maximum water vapor capacity of air is expressed as the *saturation vapor pressure* and is presented in Figure 8.1.

1. Using the graph provided in Figure 8.1, *plot* the saturation vapor pressure data in millibars and *connect* the data points with a smooth, curved line. The first several points are plotted for you.

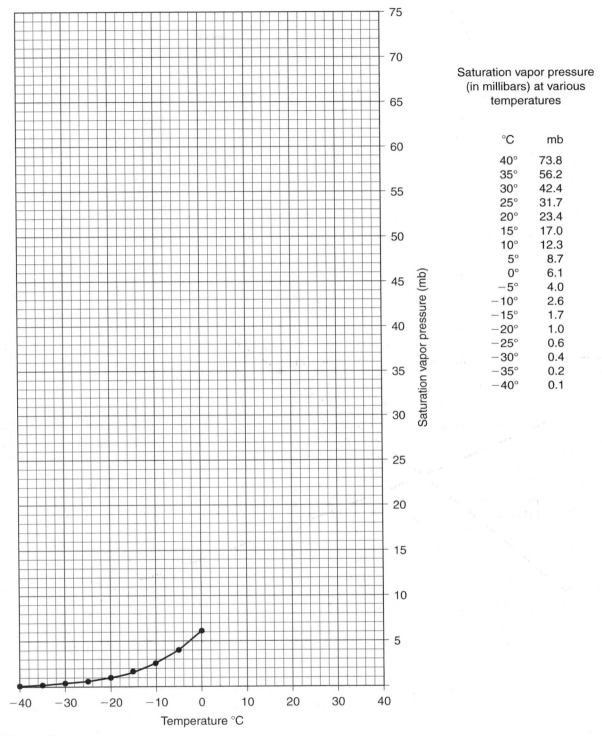

Saturation vapor pressure (in millibars) at various temperatures

°C	mb
40°	73.8
35°	56.2
30°	42.4
25°	31.7
20°	23.4
15°	17.0
10°	12.3
5°	8.7
0°	6.1
−5°	4.0
−10°	2.6
−15°	1.7
−20°	1.0
−25°	0.6
−30°	0.4
−35°	0.2
−40°	0.1

Figure 8.1
Saturation vapor pressure (mb) of air at various temperatures—maximum water vapor capacity of the air

After finishing the saturation vapor pressure graph in Figure 8.1, complete the following.

2. The difference in saturation vapor pressure between −30°C and −10°C is _____ mb; whereas

the difference in the capacity of air to hold water vapor between +10°C and +30°C is _____ mb.
Note the geometric relationship between increasing temperature and increasing capacity of the air to hold
water vapor; for each 10°C increase in temperature, the vapor pressure capacity of the air nearly doubles.

3. Air at 20°C has a saturation vapor pressure of _____ mb (*capacity*).
An air mass at 20°C actually contains water vapor exerting 12.3 mb vapor pressure (*content*). Using the
equation given at the beginning of Section 1, what is the present relative humidity of this air mass?

_____ %. Show your work: _____

_____

_____

4. What is the approximate dew-point temperature of the air mass in question #3? _____ °C? (What is
the temperature for which 12.3 mb of vapor pressure is the maximum capacity?) In other words, this air

mass must cool down from its present temperature by _____ C° to achieve
saturated conditions at the dew-point temperature. (The present temperature must cool down to the
dew-point temperature for its water *content* to equal the air's water vapor *capacity*.)

5. Using the same air mass, assume that it warms to 35°C on a hot afternoon and maintains the same vapor

pressure content of 12.3 mb. What is the relative humidity in the air mass at this time? _____ %.
(Hint: determine the capacity of 35°C air to hold water vapor, expressed as a pressure. Using the formula at

the beginning of Section 1, do the simple math.) Show your work: _____

_____

_____

6. Assuming a different air mass from the above that has a relative humidity of 40% at a temperature of 25°C,

what is the actual humidity content expressed as vapor pressure? _____ mb. Thus, by knowing
relative humidity and air temperature, relative humidity provides us with an indirect method of determining
the actual water vapor content of the air.

7. How is it possible to have an air mass over the hot, arid Sahara (30° N) contain an average of 17 mb vapor
pressure whereas an air mass in the moist midlatitudes contains an average of 8 mb vapor pressure (50° N)?
(Explain this in the context of relative humidity, average temperatures in these two regions, and water vapor

capacity of the air.) _____

_____

_____

Measuring Relative Humidity

The next step is to actually measure relative humidity. Relative humidity is measured with various instruments. The **hair hygrometer** uses the principle that human hair changes as much as 4% in length between 0% and 100% relative humidity. The instrument connects a standardized bundle of human hair through a mechanism to a gauge and a graph to indicate relative humidity. Remember: relative humidity is not a direct measurement of water vapor. Once you determine the relative humidity you can then calculate the actual water vapor content of the air.

Another instrument used to calculate relative humidity is a **sling psychrometer**. This instrument has two thermometers mounted side-by-side on a metal holder. One is called the *dry-bulb thermometer*—it simply records the *ambient* (surrounding) air temperature. The other thermometer is called the *wet-bulb thermometer*—it is set lower in the holder and has a cloth wick over the bulb that is moistened with dis-

tilled water. The psychrometer is then spun by its handle. In a weather-shelter installation, the wet-bulb thermometer can have a long cloth wick that extends to a bowl of distilled water. Instead of spinning the psychrometer, a small fan is used to move air across the dampened wick.

The rate at which water evaporates from the wick depends on the relative saturation of the surrounding air. If the air is dry, water evaporates quickly, absorbing the latent heat of evaporation from the wet-bulb thermometer, causing the temperature to drop; in other words, the wet-bulb thermometer shows a *depression* in temperature. In an area of high humidity, less water evaporates from the wick. After spinning or ventilating the psychrometer for a minute or two, the temperature on each bulb is noted and compared on a relative humidity psychrometric chart (Table 8.1) to determine relative humidity.

1. When using a sling psychrometer, what happens to the two thermometers as the psychrometer is spun? How do these changes make it possible to measure relative humidity? Explain. _____

_____

_____

Table 8.1 presents a psychrometric table that expresses dry- and wet-bulb temperature relationships in terms of *relative humidity*—the percent of water vapor actually in the air as compared to the maximum capacity that the air could hold at a given temperature. Table 8.2 presents a psychrometric table that expresses dry- and wet-bulb temperature relationships in terms of *dew-point temperatures*—that temperature at which the actual vapor pressure and the saturation vapor pressure are equal.

When using a psychrometric table, you use the wet- and dry-bulb thermometer readings as follows:

$$T = \text{Temperature of the dry bulb (air temperature)}$$

$$T_w = \text{Temperature of the wet bulb}$$

Therefore, $$T - T_w = \text{Wet-bulb depression}$$

Table 8.1
Psychrometric chart of relative humidity (in percent)

Depression of the wet bulb (dry-bulb temperature minus wet-bulb temperature in °C)

Dry-bulb temp (air temp °C)	0.5	1.0	1.5	2.0	2.5	3.0	3.5	4.0	4.5	5.0	7.5	10.0	12.5	15.0	17.5	20.0	22.5	25.0
−20	70	41	11															
−17.5	75	51	26	2														
−15	79	58	38	18														
−12.5	82	65	47	30	13													
−10	85	69	54	39	24	10												
−17.5	87	73	60	48	35	22	10											
−5	88	77	66	54	43	32	21	11	1									
−2.5	90	80	70	60	50	42	37	22	12	3								
0	91	82	73	65	56	47	39	31	23	15								
2.5	92	84	76	68	61	53	46	38	31	24								
5	93	86	78	71	65	58	51	45	38	32	1							
7.5	93	87	80	74	68	62	56	50	44	38	11							
10	94	88	82	76	71	65	60	54	49	44	19							
12.5	94	89	84	78	73	68	63	58	53	48	25	4						
15	95	90	85	80	75	70	66	61	57	52	31	12						
17.5	95	90	86	81	77	72	68	64	60	55	36	18	2					
20	95	91	87	82	78	74	70	66	62	58	40	24	8					
22.5	96	92	87	83	80	76	72	68	64	61	44	28	14	1				
25	96	92	88	84	81	77	73	70	66	63	47	32	19	7				
27.5	96	92	89	85	82	78	75	71	68	65	50	36	23	12	1			
30	96	93	89	86	82	79	76	73	70	67	52	39	27	16	6			
32.5	97	93	90	86	83	80	77	74	71	68	54	42	30	20	11	1		
35	97	93	90	87	84	81	78	75	72	69	56	44	33	23	14	6		
37.5	97	94	91	87	85	82	79	76	73	70	58	46	36	26	18	10	3	
40	97	94	91	88	85	82	79	77	74	72	59	48	38	29	21	13	6	
42.5	97	94	91	88	86	83	80	78	75	72	61	50	40	31	23	16	9	2
45	97	94	91	89	86	83	81	78	76	73	62	51	42	33	26	18	12	6
47.5	97	94	92	89	86	84	81	79	76	74	63	53	44	35	28	21	15	9
50	97	95	92	89	87	84	82	79	77	75	64	54	45	37	30	23	17	11

Table 8.2
Psychrometric chart of dew-point temperature (in °C)

Depression of the wet bulb (dry-bulb temperature minus wet-bulb temperature in °C)

Dry-bulb temp (air temp °C)	0.5	1.0	1.5	2.0	2.5	3.0	3.5	4.0	4.5	5.0	7.5	10.0	12.5	15.0	17.5	20.0
−20	−25	−33														
−17.5	−21	−27	−38													
−15	−19	−23	−28													
−12.5	−15	−18	−22	−29												
−10	−12	−14	−18	−21	−27	−36										
−5	−9	−11	−14	−17	−20	−26	−34									
−2.5	−7	−8	−10	−13	−16	−19	−24	−31								
−25	−4	−6	−7	−9	−11	−14	−17	−22	−28	−41						
0	−1	−3	−4	−6	−8	−10	−12	−15	−19	−24						
2.5	1	0	−1	−3	−4	−6	−8	−10	−13	−16						
5	4	3	2	0	−1	−2	−4	−6	−8	−10	−48					
7.5	6	6	4	3	2	1	−1	−2	−4	−6	−22					
10	9	8	7	6	5	4	2	1	0	−2	−13					
12.5	12	11	10	9	8	7	6	4	3	2	−7	−28				
15	14	13	12	12	11	10	9	8	7	5	−2	−14				
17.5	17	16	15	14	13	12	12	11	10	8	2	−7	−35			
20	19	18	18	17	16	15	14	14	13	12	6	−1	−15			
22.5	22	21	20	20	19	18	17	16	15	10	3	−6	−38			
25	24	24	23	22	21	21	20	19	18	18	13	7	0	−14		
27.5	27	26	26	25	24	23	23	22	21	20	16	11	5	−5	−32	
30	29	29	28	27	27	26	25	25	24	23	19	14	9	2	−11	
32.5	32	31	31	30	29	29	28	27	26	26	22	18	13	7	−2	
35	34	34	33	32	32	31	31	30	29	28	25	21	16	11	4	
37.5	37	36	36	35	34	34	33	32	32	31	28	24	20	15	9	0
40	39	39	38	38	37	36	36	35	34	34	30	27	23	18	13	6
42.5	42	41	41	40	40	39	38	38	37	36	33	30	26	22	17	11
45	44	44	43	43	42	42	41	40	40	39	36	33	29	25	21	15
47.5	47	46	46	45	45	44	44	43	42	42	39	35	32	28	24	19
50	49	49	48	48	47	47	46	45	45	44	41	38	35	31	28	23

2. Using the psychrometric charts in Table 8.1 and 8.2, determine the relationships asked for in Table 8.3. The first line is completed for you.

Table 8.3
Psychrometric relationships

(T) Dry-bulb temperature (°C)	(T_w) Wet-bulb temperature (°C)	(T − T_w) Wet-bulb depression (C°)	(RH) Relative humidity (%)	(T_dp) Dew-point temperature (°C)
−10°	−12°	[2°]	39%	−21°
5°	1°			
17.5°			86%	
25°	10°			
	30°			23°
37.5°	33°			32°

✳ SECTION 3

Measuring Relative Humidity and Temperature

At any given time, temperature and relative humidity may vary from one place to another, even over small distances. We have all experienced these changes in "micro-climates" as we move from one place to another. Differing local conditions can greatly influence the organisms that are present. Likewise, at a given place the relative humidity will change over time as other atmospheric conditions change. This happens every day as air temperatures rise and fall throughout a 24-hour period.

1. Your instructor will provide you with a map of your campus (or another place such as a park, etc.) on which a number of different sites have been indicated. These sites will vary in terms of whether they are in sunlight or shade, type of surface cover, and other variables. (As an option, your instructor will provide you with data already obtained from your campus or another place with which to complete this section of the exercise.)
   Record data in Table 8.4 as follows:

Table 8.4
Microclimate site characteristics and data

Site	Site characteristics and albedo	(T) Dry-bulb temperature (°C)	(T_w) Wet-bulb temperature (°C)	(T − T_w) Wet-bulb depression (C°)	(RH) Relative humidity (%)
A					
B					
C					
D					

Table 8.5
Selected albedo values of various surfaces

Surface	Albedo (% reflected)
Fresh snow	80–95
Polluted or old snow	40–70
Ocean (Sun near horizon)	40
Ocean (Sun halfway up sky)	5
Coniferous forests	10–15
Deciduous forest (green)	15–20
Crops (green)	10–25
Bare, dry sandy soil	25–45
Bare dark soil	5–15
White sand	40
Desert	25–30
Meadow, grass	15–25
Concrete	20–30
Asphalt	5–10

At each locale, record site characteristics, and from the information in Table 8.5 estimate the albedo. **Albedo**, the reflectivity of a surface, lessens absorption of Sun energy that will, in turn, mean lower surface temperatures. The lighter the surface, the higher the albedo and lower the temperature recorded. The darker the surface, the lower the albedo and higher the temperature recorded.

Following instructions given by your lab instructor, take two readings at each site with the sling psychrometer and record the temperature readings in Table 8.4. Calculate the wet-bulb depression, and determine the relative humidity from Table 8.1 in ✳ SECTION 2.

2. What variations, if any, can you observe in the recorded temperature data for the sites?

_____

_____

3. What relationship, if any, can be observed between surface albedo and temperature readings at each site?

_____

_____

4. Describe and explain the variations you measured in the relative humidity for each site.

_____

_____

_____

5. Do you feel you have identified a relation between albedo and air temperature? _____ Between

air temperature and relative humidity? _____ Briefly explain: _____

_____

_____

❋ SECTION 4

Atmospheric Stability

Knowing the concepts of humidity, relative humidity, dew-point temperature, and saturation allows us to take the next step and examine the stability of air masses. Temperature and moisture relationships in the environment determine whether the atmosphere is stable or unstable.

Stability refers to the tendency of a parcel of air either to remain as it is or to change its initial position by ascending or descending. An air parcel is termed *stable* if it resists displacement upward or, when disturbed, it tends to return to its starting place. On the other hand, an air parcel is considered *unstable* when it continues to rise until it reaches an altitude where the surrounding air has a density similar to its own.

As noted in Lab Exercise 5, ❋ SECTION 1, the **normal lapse rate** is the *average* decrease in temperature with increasing altitude, a value of 6.4 C° per 1000 m or 0.64 C° per 100 m (3.5 F° per 1000 ft). This rate of temperature change can differ greatly under varying weather conditions, and so the actual lapse rate at a particular place and time is labeled the **environmental lapse rate**. When a particular environmental lapse rate is in effect, an ascending parcel of air tends to cool by expansion, responding to the reduced pressure at higher altitudes. Descending air tends to heat by compression.

Both ascending and descending temperature changes are assumed to occur without any heat exchange between the surrounding environment and the vertically moving parcel of air. The warming and cooling rates for a parcel of expanding or compressing air are termed **adiabatic**. Adiabatic rates are measured with one of two dynamic rates, depending on moisture conditions in the parcel: **dry adiabatic rate** (DAR), 10 C° per 1000 m, or 1 C° per 100 m (5.5 F° per 1000 ft) and **moist adiabatic rate** (MAR)—sometimes also called *wet adiabatic rate (WAR)*—6 C° per 1000 m, or 0.6 C° per 100 m (3.3 F° per 1000 ft), or roughly 4 C° less than the DAR. **Note**: the DAR is used if the air is less than saturated; the MAR is used if the air is saturated.

Determining the degree of stability or instability involves measuring and comparing simple temperature relationships between conditions inside an air parcel and the surrounding air. This difference between an air parcel and the surrounding environment produces a buoyancy that contributes to further lifting. The following three problems involve each of the possible conditions: *stable* (if the rising air is cooler than or the same temperature as the surrounding air), *unstable* (if the rising air is warmer than the surrounding air), and *conditionally unstable* (if, in rising, it becomes warmer than the surrounding air only after it begins cooling at the MAR). The basic temperature relationships that determine stability conditions in the atmosphere between the environmental lapse rate, and dry and moist adiabatic rates are shown in Figure 8.2.

For clarity, use color pencils to highlight each of the three lapse rates in Figure 8.2. The normal lapse rate (remember: it's an average), the DAR, and MAR rates are plotted according to their respective formulas. The curved set of arrows shows the overall possible range within which the actual environmental rate might fall. If the environmental lapse rate of the air surrounding a rising parcel of air is greater than the DAR (i.e., >10 C°/1000 m), the rising air will remain warmer than the surrounding environment, will be *unstable* (a), and will continue to rise. If the environmental lapse rate is less than both the DAR and the MAR (i.e., < 6 C°/1000 m), the rising column of air will be cooler than the surrounding air (which is cooling at the environmental rate) and will be *stable* (c), resisting further rising, and instead will settle back to its original position. In the circumstance that the environmental rate is between the MAR and DAR (>MAR and <DAR), the parcel of rising air will resist rising as long as the air remains unsaturated; however, if the parcel is forced to rise (over a mountain range, for example) and the air becomes saturated, it will then begin cooling at the MAR, remaining warmer than the surrounding air, and become unstable, continuing to rise (b). This is *conditionally unstable* air: unstable only under certain conditions—in this instance, being forced to rise.

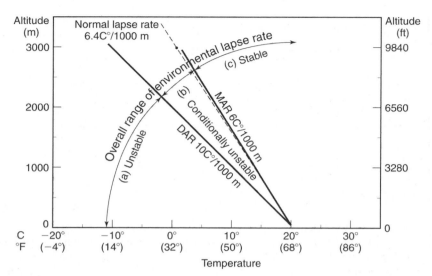

Figure 8.2
The relationship between dry and moist adiabatic rates and environmental lapse rates produces three conditions of stability.

1. On the graphs provided in Figure 8.3, *plot* the following data, using color pencils to distinguish the different rates. Begin with a surface temperature of 25°C noted on all three graphs labeled a, b, and c. Please use color: ELR = green pencil; DAR = red pencil; MAR = blue pencil.

   a) Plot an environmental lapse rate of 11 C° per 1000 m, DAR of 10 C° per 1000 m, MAR of 6 C° per 1000 m. This graph is completed for you.

   b) Plot an environmental lapse rate of 4.5 C° per 1000 m, DAR of 10 C° per 1000 m, MAR of 6 C° per 1000 m.

   c) Plot an environmental lapse rate of 8 C° per 1000 m, DAR of 10 C° per 1000 m, MAR of 6 C° per 1000 m.

2. Using Figure 8.2 and your graphs in Figure 8.3, note the relationship between the environmental lapse rates on each graph you made and the DAR and MAR. Determine which stability condition describes each of the three graphs you have completed. Hint: compare each ELR plotted with the relations in Figure 8.2.

   Stability graph (a) _____

   Stability graph (b) _____

   Stability graph (c) _____

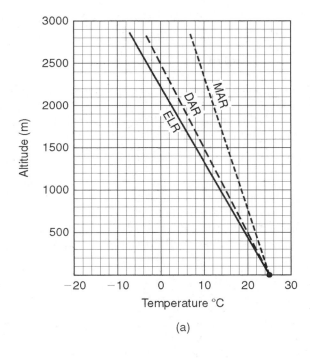

(a)

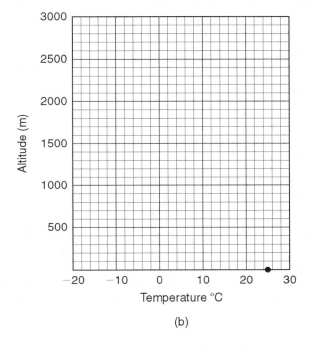

(b)

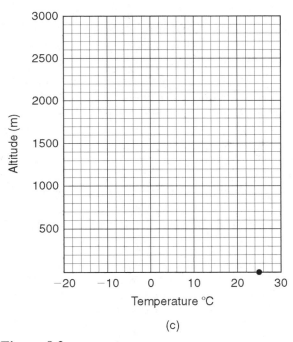

(c)

Figure 8.3
Three atmospheric stability graphs

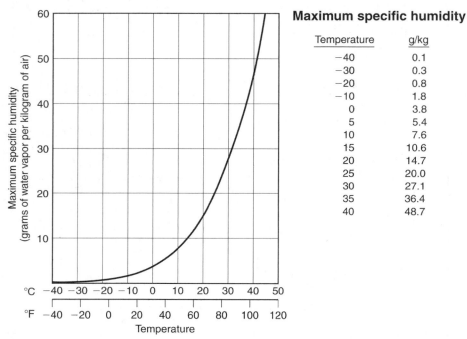

✳ SECTION 5

An Example of Unstable Atmospheric Conditions

Let's work through a specific application of these stability principles. In this example, **specific humidity** is used as another way of describing water vapor content. Specific humidity describes the water vapor content of the air by comparing a mass of water vapor to a mass of air, expressed as grams per kilogram. Figure 8.4 presents the *maximum specific humidity* for a mass of air at various temperatures.

The graph in Figure 8.4 shows that a kilogram of air could hold a maximum specific humidity of 47 g of water vapor at 40°C (104°F), 15 g at 20°C (68°F), and 4 g at 0°C (32°F). *Locate* these three data points on the graph and use a color pencil to *mark* them (with a dot).

If a kilogram of air at 40°C has a specific humidity of 12 g, its relative humidity is calculated as follows:

$$12 \text{ g} \div 47 \text{ g} = 0.255 \times 100 = 25.5\%$$

The actual water vapor in the air (12 g) divided by the maximum specific humidity at 40°C (47 g), gives you the relative humidity. Specific humidity is useful in describing the moisture content of large air masses that are interacting in a weather system.

Maximum specific humidity

Temperature	g/kg
−40	0.1
−30	0.3
−20	0.8
−10	1.8
0	3.8
5	5.4
10	7.6
15	10.6
20	14.7
25	20.0
30	27.1
35	36.4
40	48.7

Figure 8.4
Maximum specific humidity of air at various temperatures–maximum water vapor capacity of the air

1. Assume that an air parcel has a specific humidity of 10 g/kg and an internal temperature of 25°C. The parcel of air begins to lift. Assume an *environmental lapse rate* of 12 C° in the air surrounding the lifting parcel. Use standard values for DAR and MAR given in ✳ SECTION 4. Using the graph provided in Figure 8.5 (below), *plot* and label the environmental lapse rate. (Refer to the stability graph you plotted earlier in Figure 8.3a.)

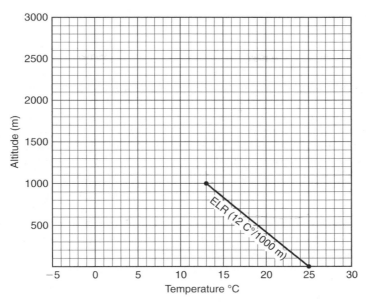

Figure 8.5
Unstable atmospheric conditions and convective activity.

2. Determine the *dew-point temperature* at which the rising air will become saturated (see Figure 8.4 for

   maximum specific humidity at various temperatures): _____

   a) Plot the DAR on the graph for the lifting parcel of air until it cools to the dew-point temperature.
   b) Label the DAR and the dew-point temperature.
   c) Continue plotting from the dew-point mark, but now use the slower cooling MAR to the top of the graph; label the MAR. (Again, refer to Figure 8.2a for help.)

3. At what altitude does the lifting air parcel achieve the dew-point temperature and the **lifting condensation level** (the altitude at which 100% relative humidity—*saturation*—is reached and the MAR begins)

   given these conditions? _____

   Show your work. _____

   Draw and label a horizontal line on the graph marking this lifting condensation level.

4. Is the air parcel unstable (continued lifting) at 3000 m altitude? _____

   Explain. _____

   _____

   _____

5. According to a physical geography text, what two prerequisites must exist for the formation of clouds to

   occur in the air parcel when it reaches its dew-point temperature? Explain. _____

   _____

   _____

6. At what altitude would you expect clouds to form in Figure 8.5?

Why? _____

_____

_____

7. In the margin next to the graph in Figure 8.5, sketch the lifting parcel of air (use a dotted line to suggest its transparency) and draw a cumulus cloud formation at the appropriate altitude where condensation begins.

_____

8. As the parcel in Figure 8.5 rises to 3000 m, it will cool. Under normal conditions relative humidity does not exceed 100%. If the parcel rises to 3000 m, what is the difference in specific humidity expressed in grams per

kilogram between the parcel's initial conditions on the ground and at 3000 m? _____

If the parcel descends, how would its relative humidity change? How does this relate to the rain shadow concept?

_____

Note: There are four principal lifting mechanisms that can force air to rise, resulting in adiabatic cooling, and perhaps reach its dew-point temperature. The same stability concepts you sketched and worked with in Figure 8.5 apply in each of these lifting situations despite the difference in mechanism. These lifting mechanisms are:

a) *convergent* (air flowing from different directions into an area of low pressure);

b) *convectional* (local heating of air, a plowed field, an urban heat island, land compared to water);

c) *orographic* (air encounters a mountain barrier and is forced to lift); and,

d) *frontal* (conflicting air masses of different densities forcing warmer air to lift).

✳ SECTION 6

Orographic Lifting and Rain Shadows

The physical presence of a mountain acts as a topographic barrier to migrating air masses. Orographic lifting (oro means "mountain") occurs when air is forcibly lifted upslope as it is pushed against a mountain. The lifting air cools adiabatically. An orographic barrier enhances convectional activity and causes additional lifting during the passage of weather fronts and cyclonic systems, thereby extracting more moisture from passing air masses. The wetter intercepting slope is the windward slope, as opposed to the drier far-side slope,

known as the leeward slope. Figure 8.6 on the next page is a transect across the Sierra Nevada mountains in California. We will follow a storm across them from Fresno up to Mt. Whitney, the highest point in the continental United States, down to Lone Pine, back up to Telescope Peak and finally down to Badwater in Death Valley, the lowest point in North America. When the parcel is descending assume the air is stable and descends to the surface. Use Figure 8.4 to calculate the maximum specific humidity values for each elevation.

1. What is the condensation height for this storm? _____

2. Fill in the answer blanks in Figure 8.6.

3. Mt. Whitney

4421 m

Temperature: _____

Specific humidity: _____

Relative humidity: _____

Grams of H$_2$O lost
since last station: _____

5. Telescope Peak

3368 m

Temperature: _____

Specific humidity: _____

Relative humidity: _____

Grams of H$_2$O lost
since last station: _____

6. Death Valley

−82 m

Temperature: _____

Specific humidity: _____

Relative humidity: _____

Grams of H$_2$O lost
since last station: _____

2. King's Canyon

2000 m

Temperature: _____

Specific humidity: _____

Relative humidity: _____

Grams of H$_2$O lost
since last station: _____

4. Lone Pine

1130 m

Temperature: _____

Specific Humidity: _____

Relative humidity: _____

Grams of H$_2$O lost
since last station: _____

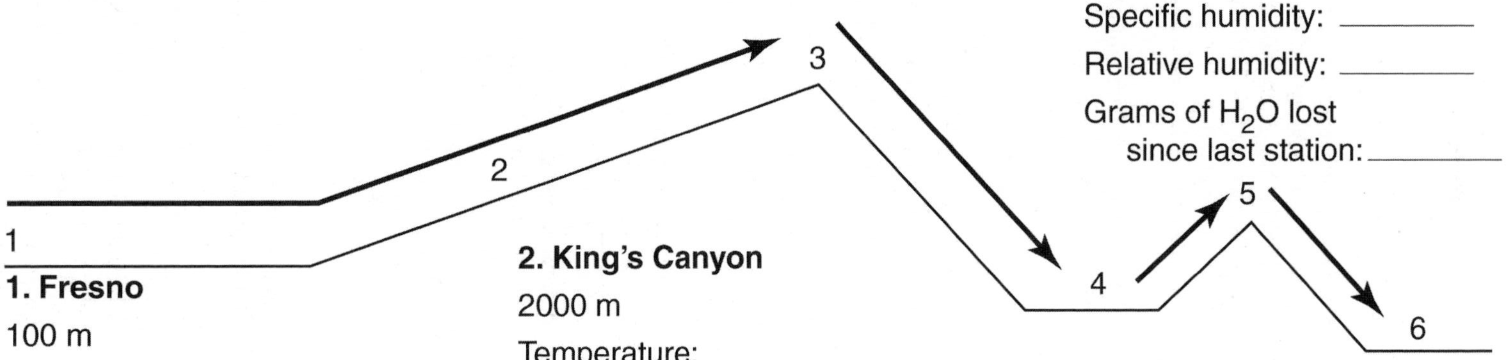

1. Fresno

100 m

Temperature: _____10°C_____

Specific humidity: _____

Relative humidity: _____50%_____

Grams of H$_2$O lost
since last station: _____0_____

Figure 8.6

3. Do you live on the windward or leeward (rain shadow) side of the nearest major terrain feature? What is the nearest terrain feature that controls your precipitation?

_____

_____

The Indian Ocean is very warm and can reach temperatures of 30°C. This warm body of water, combined with low air pressure in central Asia, creates monsoonal conditions during the summer. The low air pressure pulls warm moist air from the Indian Ocean up the slopes of the foothills and main body of the Himalayas.

Cherrapunji, India is one of the wettest places in the world and is nestled in the foothills in this region. Cherrapunji holds the record for the wettest year ever recorded an amazing 26.47 m (86.8 ft), although it sounds even more impressive when given in the normal rainfall reporting units of 2,647 mm or 1,042 in.!

4. Take a parcel of air from the Bay of Bengal in the Indian Ocean up to Cherrapunji at 1,313 m (4309 ft). Assume its temperature was 30°C and 90% RH at 0 m (0 ft) over the Bay of Bengal. What would its initial

SH be? _____

What would its SH be at 1,313 m (4,309 ft)? _____

How many mm of H_2O would fall as rain along the way (assume that any moisture in excess of the maximum specific humidity will fall as rain and that 1g of H_2O is equal to 10 mm of precipitation)?

_____

What would the parcel's condensation height be? _____

_____

_____

While moisture condenses from the lifting air mass on the windward side of the mountain; on the leeward side, the descending air mass heats by compression, and any remaining liquid water in the air evaporates. The previous questions looked at air parcels beginning their ascent while the air was warm and moist, but this question looks at an air parcel when it is finishing its descent on the leeward slope and it becomes hot and

dry. As stable air descends it is heated by compression at the dry adiabatic rate (DAR) of 10 C°/km (5.5 F° per 1000 ft). This heating causes the relative humidity to drop, causing warm dry winds. These winds go by a number of regional names including *foehn* or *föhn* in Europe and Chinook or snow eater in the Pacific North West in tribute to their ability to melt away snow.

5. Take a parcel of air from the top of Mt. Rainier 4,392 m (14,410 ft) and follow it to Yakima, WA 325 m (1,066 ft). Assume the initial conditions of the parcel of air are −10°C and 100% RH.

What is the initial specific humidity of the parcel? _____

What will its temperature, specific humidity and relative humidity be in Yakima? _____

_____

How many degrees warmer is the parcel in Yakima than on top of Mt. Rainier? _____

Name: _____   Laboratory Section:_____

Date: _____   Score/Grade: _____

Lab Exercise 9

Weather Maps

Weather is the short-term, day-to-day condition of the atmosphere, contrasted with **climate**, which is the long-term average (over decades) of weather conditions and extremes in a region. Weather is, at the same time, both a "snapshot" of atmospheric conditions and a technical status report of the Earth-atmosphere heat-energy budget.

We consult the broadcast media for weather information so that we can better plan our activities and travels. We turn for the weather forecast to the National Weather Service in the United States (**http://www.nws.noaa.gov**) or Canadian Meteorological Centre, a branch of the Meteorological Service of Canada (MSC) (**http://weatheroffice.ec.gc.ca/canada_e.html**) to see current satellite images and to hear weather analysis. Internationally, the World Meteorological Organization coordinates weather information (see **http://www.wmo.ch/**). Perhaps, you are already working on the "Weather Calendar" assignment, gathering weather data and recording it in the proper form and with the official symbols used internationally.

In this exercise, we work with an actual weather map to better understand this important aspect of our daily lives. Lab Exercise 9 features three sections.

You have the option of constructing an actual weather map in ❈ SECTION 2 or analyzing an idealized weather map in ❈ SECTION 3, or perhaps, completing both as directed by your instructor.

The final section examines the 2005 Atlantic **hurricane** season. Students will plot the locations of hurricane Katrina, the most economically damaging hurricane ever recorded in the Atlantic. By analyzing the relationship between hurricane behavior and sea surface temperatures, we will examine the role of hurricanes in the Earth-atmosphere heat-energy budget. After examining the physical inputs into hurricanes, we will consider the human-environment relationship.

Key Terms and Concepts

climate
latent heat
meteorology
synoptic map

weather
hurricane
tropical cyclone
typhoon

Objectives

After completion of this lab, you should be able to:

1. *Recognize* the symbols used to depict weather conditions at stations on weather maps.
2. *Construct* an actual weather map by *drawing* isobars, fronts, and *labeling* air masses, and then *portray* weather conditions experienced on that day.
3. *Analyze* an idealized weather map and *describe* current conditions at various locations.
4. *Plot* hurricane locations, evaluate the effects of sea surface temperatures on wind speeds and air pressure, and predict potential effects of climate change on hurricane behavior.

Materials/Sources Needed

pencil
colored pencils

Lab Exercise and Activities

❋ SECTION 1

Weather Map Symbols

A "Weather Calendar" exercise is assigned in the Prologue Lab. If you are working on the calendar exercise,
then you are now familiar with the standard weather
station symbol and the shorthand method used by meteorologists to describe synoptic conditions. Remember: *weather* is the short-term condition of the
atmosphere, as compared to *climate*, which reflects
long-term atmospheric conditions and extremes.

   Meteorology is the scientific study of the atmosphere (*meteor* means "heavenly" or "of the atmosphere"). Embodied within this science is a study of the
atmosphere's physical characteristics and motions;
related chemical, physical, and geological processes;
the complex linkages of atmospheric systems; and
weather forecasting.

   The Internet and its World Wide Web are a rich
source of weather information. Over 30 sources are listed
under "Destinations" in Chapter 8 of our *Geosystems*, or
Chapter 5 for *Elemental Geosystems*, Home Page. The
URL for our web site is **http://www.prenhall.com/
christopherson**.

   The sample station model is shown in Figure 9.1;
a simplified version of the station model appropriate
for use here is presented in the Prologue Lab. Note the
variety and complexity of weather conditions noted in
this simple sample station model. Figure 9.1 includes
symbol explanations for present weather conditions,
state of the sky—sky coverage, cloud designations, air
pressure changes, wind indicators and speeds, and
surface front designations.

   Sample weather station recording using standard weather symbols.

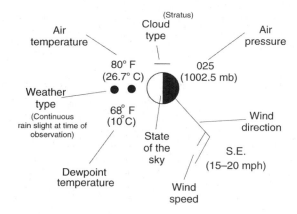

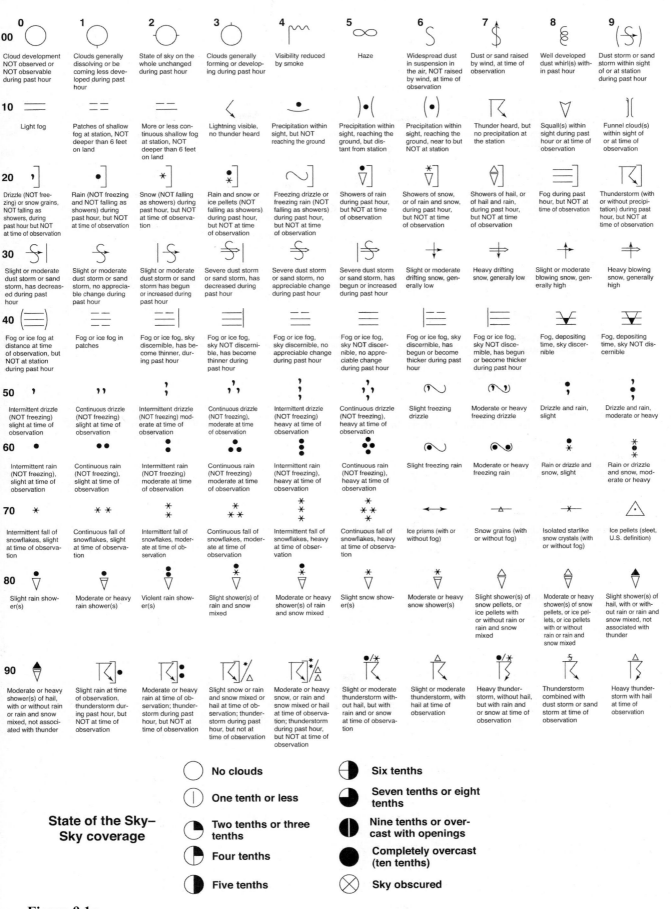

Figure 9.1a

Sample station reporting symbols and codes (2 pages)

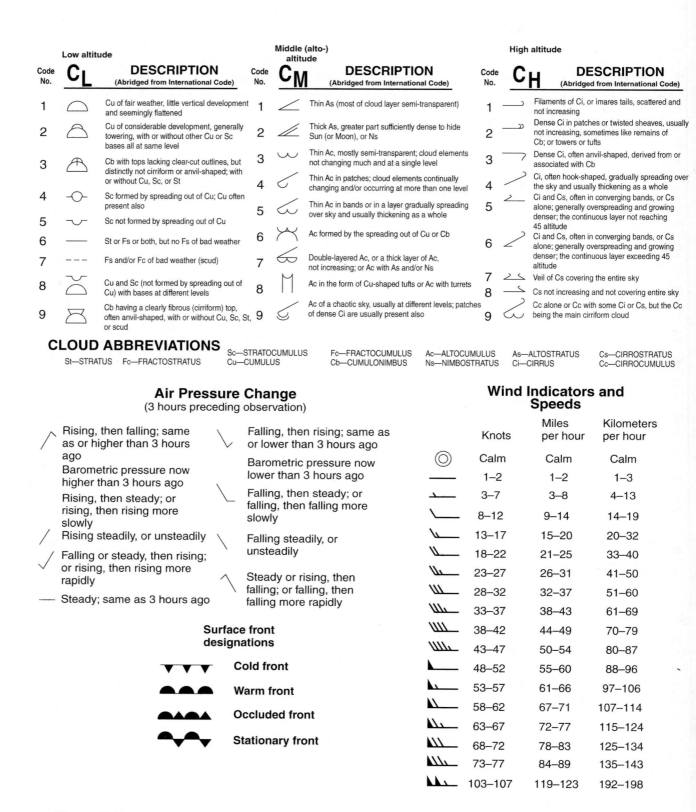

Code No.	C_L	DESCRIPTION (Abridged from International Code)
1		Cu of fair weather, little vertical development and seemingly flattened
2		Cu of considerable development, generally towering, with or without other Cu or Sc bases all at same level
3		Cb with tops lacking clear-cut outlines, but distinctly not cirriform or anvil-shaped; with or without Cu, Sc, or St
4		Sc formed by spreading out of Cu; Cu often present also
5		Sc not formed by spreading out of Cu
6		St or Fs or both, but no Fs of bad weather
7		Fs and/or Fc of bad weather (scud)
8		Cu and Sc (not formed by spreading out of Cu) with bases at different levels
9		Cb having a clearly fibrous (cirriform) top, often anvil-shaped, with or without Cu, Sc, St, or scud

Middle (alto-) altitude

Code No.	C_M	DESCRIPTION (Abridged from International Code)
1		Thin As (most of cloud layer semi-transparent)
2		Thick As, greater part sufficiently dense to hide Sun (or Moon), or Ns
3		Thin Ac, mostly semi-transparent; cloud elements not changing much and at a single level
4		Thin Ac in patches; cloud elements continually changing and/or occurring at more than one level
5		Thin Ac in bands or in a layer gradually spreading over sky and usually thickening as a whole
6		Ac formed by the spreading out of Cu or Cb
7		Double-layered Ac, or a thick layer of Ac, not increasing; or Ac with As and/or Ns
8		Ac in the form of Cu-shaped tufts or Ac with turrets
9		Ac of a chaotic sky, usually at different levels; patches of dense Ci are usually present also

High altitude

Code No.	C_H	DESCRIPTION (Abridged from International Code)
1		Filaments of Ci, or imares tails, scattered and not increasing
2		Dense Ci in patches or twisted sheaves, usually not increasing, sometimes like remains of Cb; or towers or tufts
3		Dense Ci, often anvil-shaped, derived from or associated with Cb
4		Ci, often hook-shaped, gradually spreading over the sky and usually thickening as a whole
5		Ci and Cs, often in converging bands, or Cs alone; generally overspreading and growing denser; the continuous layer not reaching 45 altitude
6		Ci and Cs, often in converging bands, or Cs alone; generally overspreading and growing denser; the continuous layer exceeding 45 altitude
7		Veil of Cs covering the entire sky
8		Cs not increasing and not covering entire sky
9		Cc alone or Cc with some Ci or Cs, but the Cc being the main cirriform cloud

CLOUD ABBREVIATIONS

St—STRATUS   Fc—FRACTOSTRATUS   Sc—STRATOCUMULUS   Cu—CUMULUS   Fc—FRACTOCUMULUS   Cb—CUMULONIMBUS   Ac—ALTOCUMULUS   Ns—NIMBOSTRATUS   As—ALTOSTRATUS   Ci—CIRRUS   Cs—CIRROSTRATUS   Cc—CIRROCUMULUS

Air Pressure Change
(3 hours preceding observation)

Rising, then falling; same as or higher than 3 hours ago

Barometric pressure now higher than 3 hours ago

Rising, then steady; or rising, then rising more slowly

Rising steadily, or unsteadily

Falling or steady, then rising; or rising, then rising more rapidly

Steady; same as 3 hours ago

Falling, then rising; same as or lower than 3 hours ago

Barometric pressure now lower than 3 hours ago

Falling, then steady; or falling, then falling more slowly

Falling steadily, or unsteadily

Steady or rising, then falling; or falling, then falling more rapidly

Surface front designations

Cold front

Warm front

Occluded front

Stationary front

Wind Indicators and Speeds

	Knots	Miles per hour	Kilometers per hour
	Calm	Calm	Calm
	1–2	1–2	1–3
	3–7	3–8	4–13
	8–12	9–14	14–19
	13–17	15–20	20–32
	18–22	21–25	33–40
	23–27	26–31	41–50
	28–32	32–37	51–60
	33–37	38–43	61–69
	38–42	44–49	70–79
	43–47	50–54	80–87
	48–52	55–60	88–96
	53–57	61–66	97–106
	58–62	67–71	107–114
	63–67	72–77	115–124
	68–72	78–83	125–134
	73–77	84–89	135–143
	103–107	119–123	192–198

Figure 9.1b

1. Select one day from your Prologue Lab "Weather Calendar," or use the current conditions outside and record the information here in the appropriate format (shown with the model at the beginning of ✳ SECTION 1). Make notations in the proper form only for the following atmospheric conditions:

   • Air temperature (°C or °F)
   • Air pressure (inches or millibars)
   • State of the sky (sky coverage)
   • Weather type (if any observed)
   • Wind direction (if any observed)
   • Wind speed (if any observed)
   • Cloud type (if any observed)

Date_____

Place_____

Describe the weather conditions depicted for this day:

_____

_____

_____

_____

2. Record the proper symbols for each of the following listed in Figure 9.1.

   a) Visibility reduced by smoke: _____

   b) Intermittent drizzle (not freezing) heavy at time of observation: _____

   c) Ice pellets (sleet-U. S. definition): _____

   d) Fog or ice fog at a distance: _____

   e) Dust storm within sight: _____

   f) Forty-percent sky coverage: _____

   g) Double-layered altocumulus: _____

   h) 55–60 mph (88–96 kmph) winds: _____

This system of data recording is used in the preparation of the synoptic weather map such as the one you work with in ✳ SECTION 2.

Daily Weather Map—Typical April Pattern

Daily weather maps are known as **synoptic maps**, meaning that they show atmospheric conditions at a specific time and place. The daily weather map is a key analytical tool for meteorologists. Figure 9.2 presents an adapted version of the synoptic map for a typical April morning, 7:00 A.M E.S.T. This exercise involves adding appropriate isobars, fronts, and air mass designations to the map. The following discussion takes you step-by-step through the completion of this map. Follow this sequence:

1. Each circle on the map represents a weather station. The number to the upper right of the station is the barometric pressure presented in an abbreviated form used by the National Weather Service (again, reference the station model in Section 1). As an example: if a *145* appears, it is short for 1014.5 mb (the "10" and the decimal point are dropped); a *980* is 998.0 mb (the "9" and the decimal point are dropped).

   a) On April 1, 1971, the center of low pressure was near Wausau, Wisconsin (see the **L** on the map), with a pressure of 994.7 mb (947 on the map). Note the wind flags around this center of low pressure. Do the wind flags at the various stations show counterclockwise winds, as you expect in a midlatitude cyclonic circulation system? _____

   b) The high-pressure center is located near Salmon, Idaho (see the **H** on the map), with a pressure of 1033.6 mb (336 on the map). Note the pattern of air temperatures and even lower dew-point temperatures associated with a cold-air mass centered in the region of high pressure.

2. On the weather map, draw in the isobars connecting points of equal barometric pressure, at 4 mb intervals.

   a) The **996 mb** and **1000 mb** isobar around the low-pressure center is drawn for you (pressures lower than 996 go inside the isobar, higher than 996 go outside this closed isobar; pressures between 996 mb and 1000 mb fall between the two isobars).

   b) Now draw in order of increasing pressure the rest of the isobars at 4 mb intervals: **1004, 1008, 1012, 1016, 1020, 1024,** and **1028.** The highest value isobar of **1032 millibars** is drawn for you. Note the 1032 mb isobar is a closed isobar that surrounds Salmon, Idaho, which is the area of highest pressure.

3. After completion of the isobars, determine the pattern of low-pressure and high-pressure systems on the map and related air masses to determine the position of the weather fronts. The southeastern portion of the country is influenced by a mild maritime tropical (mT) air mass. The high-pressure area around Idaho is under the influence of a continental polar (cP) air mass. Note the higher temperatures and the southerly winds moving northward toward the warm front that stretches from the center of low pressure in Wisconsin, across Michigan, and into Pennsylvania.

   a) Sketch in the warm front using the proper warm-front symbol (the warm front line is started for you).

   b) To draw the cold front and leading edge of the cold air mass, locate stations with northwesterly winds and lower temperatures (to the northwest of the front's location (for instance, Missouri, Kansas, and Nebraska). Now locate cities with southerly and southwesterly winds and higher temperatures (for instance, compare northeastern Missouri against southwestern Indiana, or Arkansas against Oklahoma).

   c) Draw the cold front between these contrasting pairs of stations southwestward from Wisconsin to Texas (the cold front line is started for you).

4. On this weather map note air *temperatures* in comparison with *dew-point temperatures.* The air temperature and dew-point temperatures are noted to the left of each station symbol on the map.

   a) List a sampling of air temperatures and dew-point temperatures for Idaho, Wyoming, and Utah. Try to use specific city names (use an atlas or map):

   _____

   _____

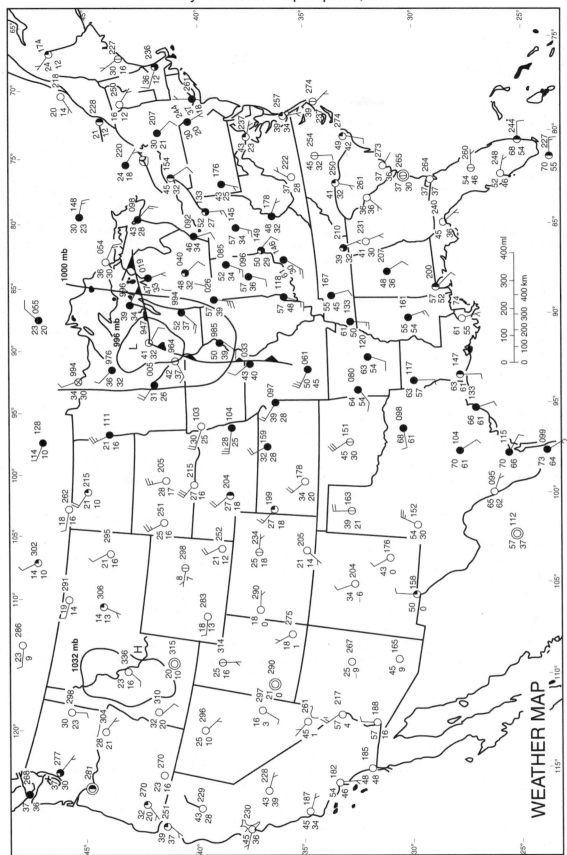

Figure 9.2
Daily weather map

b) Now compare these with <u>air temperature</u> and <u>dew-point temperatures</u> in Louisiana and Mississippi. Try to use specific city names (use an atlas or map):

_____

_____

_____

Recall from Lab Exercise 8 that the relationship between air temperature and dew-point temperature gives you an idea of the moisture content of the cP (dry) and mT (moist) air masses, respectively.

5. Describe the pattern of cloudiness across the map. The pattern of cloudiness is indicated by the state-of-the-sky (sky coverage) status recorded within each station symbol. The areas of frontal lifting are clearly identified by these patterns of clouds.

_____

_____

_____

✳ SECTION 3

Idealized Weather Map Analysis

Analyze the idealized weather map below to *determine* general weather conditions at the stations noted. Using the weather map symbols listed earlier in this lab exercise, *place* a weather symbol and information at each of the six city locations noted. Include: wind direction and speed (if any), air pressure, air temperature, dew-point temperature, state of the sky, and a guess at weather type (if any)—all approximate, generalized estimates.

Using the weather symbols presented earlier, and a physical geography text dealing with air masses and weather, complete the following information for six cities.

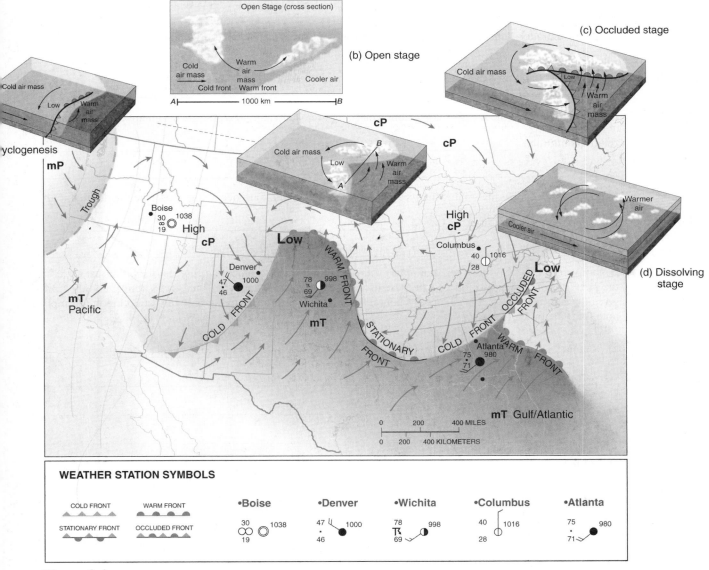

Figure 9.4

Idealized weather map—assume afternoon conditions

1. Tallahassee, Florida—label the weather conditions you think are occurring. Describe the dominant air mass and relative humidity. _____

   _____

2. Columbus, Ohio—label the weather conditions you think are occurring. Describe the dominant air mass and relative humidity. _____

   _____

3. Wichita, Kansas—label the weather conditions you think are occurring. Describe the dominant air mass and relative humidity. _____

   _____

4. Denver, Colorado—label the weather conditions you think are occurring. Describe the dominant air mass and relative humidity. _____

   _____

5. Boise, Idaho—label the weather conditions you think are occurring. Describe the dominant air mass and relative humidity. _____

   _____

6. San Francisco, California—label the weather conditions you think are occurring. Describe the dominant air mass and relative humidity. _____

   _____

7. If you were a weather forecaster in *Wichita, Kansas*, and needed to prepare a forecast for broadcast covering the next 24 hours, what would you say? Begin with current conditions you listed in #3 above, and then progress through the next 6, 12, 24, and 48 hours. Assume the system is moving eastward at 25 kmph (15 mph), note the scale on the map.

   a) 6 hours: _____

   _____

   b) 12 hours: _____

   _____

   c) 24 hours: _____

   _____

Hurricanes, Typhoons, and Cyclones

Tropical cyclones are potentially the most destructive storms experienced by humans, claiming thousands of lives each year worldwide. This is especially true when they attain the wind speeds and low-pressure readings of a full-fledged **hurricane, typhoon**, or *cyclone* (> 65 knots, > 74 mph, > 119 kmph).

The 2005 Atlantic hurricane season saw many broken records and historic firsts, including the most named tropical storms in a single year; the most hurricanes in a single year; the first time that two category 5 storms were in the Gulf of Mexico in the same year; the most intense hurricane ever recorded in the Atlantic (Hurricane Wilma); and finally, more damage was recorded in 2005, at $130 billion, than any other year. 2005 was an exceptional year, yet it is part of a larger picture worldwide.

1995–2006, has been the most active 12 year period in the history of the National Hurricane Center, with record levels of activity and individual storm intensity. Some of the other extraordinary events during this time period include:

- In March 2004, Hurricane Catarina was the first hurricane observed turning from the equator into the south Atlantic.

- In October 2005, Tropical Storm Vince became the first Atlantic tropical cyclone to strike Spain!

- In 2006, Super Typhoon Ioke, the first category 5 super typhoon, developed south of Hawai'i and lasted for almost three weeks with its final remnants causing erosion along the shores of Alaska!

- And in 2007, Super Cyclone Gonu became the strongest tropical cyclone on record occurring in the Arabian Sea, eventually hitting Oman and the Arabian Peninsula.

These four storms are examples of the increased occurrence and intensity of cyclonic systems which are related to higher oceanic and atmospheric temperatures.

Tropical cyclones are powerful manifestations of the Earth–atmosphere energy budget. The warm air and seas where they begin provide abundant water vapor and thus the necessary **latent heat** to fuel these storms. Tropical cyclones convert heat energy from the ocean into mechanical energy in the wind—the warmer the ocean and atmosphere, the more intense the conversion and powerful the storm. For these processes, sea-surface temperatures must exceed approximately 26°C (79°F). During Hurricane Katrina, the sea surface temperature (SST) in the Gulf of Mexico exceeded 30°C (86°F).

We appear to be in a more intense hurricane cycle similar to the 1941 to 1970 pattern when there were 24 major hurricanes, as compared to 14 between 1971 and 1994. Warmer Atlantic waters along the 20th parallel between 20°W and 90°W, and the Caribbean and Gulf of Mexico, is fueling a greater intensity than previously experienced. A study published in *Science* stated,

> . . . the trend of increasing numbers of category 4 and 5 hurricanes for the period 1970-2004 is directly linked to the trend in sea-surface temperature. . . . Higher SST [sea-surface temperatures] was the only statistically significant controlling variable related to the upward trend in global hurricane strength since 1970.*

Statistically, the damage caused by tropical cyclones is increasing substantially as more and more development occurs along susceptible coastlines, whereas loss of life is decreasing in most parts of the world as a result of better forecasting of these storms.

The combination of more intense and numerous storms, due in part to higher SSTs, and increasing development of inappropriate areas will create the potential for future catastrophes. Hurricane Katrina was a moderate category 3 storm, but the ensuing devastation and loss of life in the flood disaster was due to human causes, primarily the failure of protective levee systems.

Plot Katrina's storm track using the data in Figure 9.5, then fill in the sea-surface temperature (SST) for each of the data points in the spaces provided. Finally, complete the questions on p. 150.

---

*C.D. Hoyos, P.A. Agudelo, P.J. Webster, and J.A. Curry, "Deconvolution of Factors Contributing to the Increase in Global Hurricane Intensity, *Science*, (April 7, 2006), v 312 no. 5770: pp. 94–97.

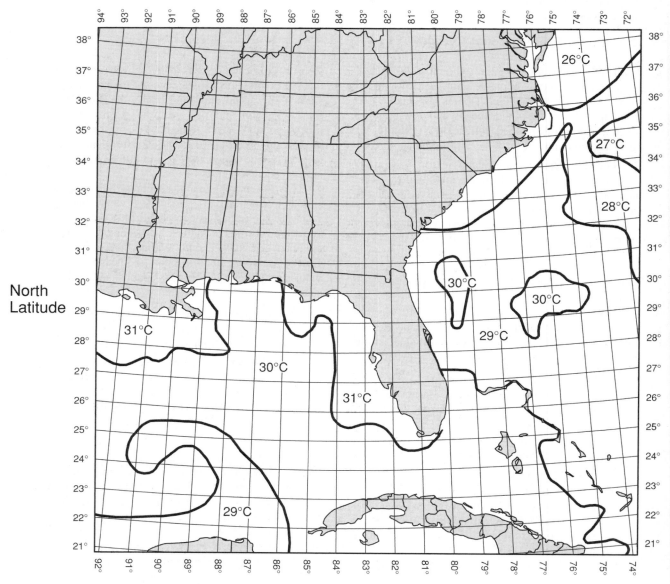

Hurricane Katrina

DateTime(UTC)	Latitude	Longitude	Mb	KMH	Category	SST
8/23/2005 18:00	23.10	75.10	1008	56	Tropical Depression	
8/24/2005 18:00	25.40	76.90	1003	74	Tropical Storm	
8/25/2005 12:00	26.20	79.60	988	111	Tropical Storm	
8/26/2005 18:00	24.90	82.60	968	157	2	
8/27/2005 18:00	24.50	85.30	948	185	3	
8/28/2005 18:00	26.30	88.60	902	278	5	
8/29/2005 12:00	29.50	89.60	923	204	3	
8/30/2005 18:00	37.00	87.00	990	30	Tropical Depression	

Figure 9.5
Hurricane Katrina

Hurricane Katrina

1. What are the physical inputs required to create a hurricane? Identify and describe the positive feedback loop that allows them to mature. What factors act to decrease hurricane intensity and strength? _____

   _____

   _____

2. How would you characterize the speed and direction of Katrina's movement from the 23rd through the 26th compared to the 27th to the 30th?

   _____

   _____

3. Estimate the distance travelled using 96 km (60 mi)/ degree of longitude and 111 km (69 mi)/ degree of latitude. How far did Katrina travel each day from the 23rd through the 26th in degrees? in km? in miles?

   _____

   How far did it travel each day from the 27th to the 30th in degrees? in km? in miles? _____

   _____

4. What was Katrina's maximum recorded wind speed? How fast was it travelling on that day in kmh? What was the maximum wind speed in the right-front quadrant (wind speed combined with travelling speed)? What would the total wind speed have been in the left-rear quadrant (wind speed minus travelling speed)? _____

   _____

   _____

5. Mark the location on the map with lowest Mb/highest wind speed. What day did that occur on? What was the water temperature on that day? _____

   _____

6. How much did the central pressure drop from normal sea-level pressure? _____

7. Fill in the table entries for SST for each day. For hurricanes to form the minimum water temperature is 26°C (79° F). What was the warmest water that the hurricane passed over? coldest water? Do you think that there is a relationship between the water temperature and wind speed? _____

   _____

8. Briefly summarize the relationship between SST (sea surface temperature) and hurricane strength. What are the implications of this on future hurricanes? What advice would you give to coastal planners regarding future development of coastal areas on the east coast? _____

   _____

   _____

Lab Exercise 10

Water Balance and Water Resources

Because water is not always naturally available when and where it is needed, humans must rearrange water resources in time and space. The maintenance of a house plant, the distribution of local water supplies, an irrigation program on a farm, the rearrangement of river flows—all involve aspects of the water balance and water-resource management.

The water-balance is a portrait of the hydrologic cycle at a specific site or area for any period of time. The water-balance can be used to estimate streamflow, accurately determine irrigation quantity and timing, and to show the relationship between a given supply of water and the local demand.

A water balance can be established for any defined area of Earth's surface—a continent, nation, region, or field—by calculating the total precipitation input and the total water output.

In this lab exercise, we work with a water-balance equation and accounting procedure to determine moisture conditions for two cities—Kingsport, Tennessee, and Sacramento, California. Given this data you will prepare graphs that illustrate these water balance relationships. Also, this lab examines the broader issues of water resources in the United States. Lab Exercise 10 features 5 sections.

Key Terms and Concepts

actual evapotranspiration
available water
capillary water
consumptive uses
deficit
evaporation
evapotranspiration
field capacity
potential evapotranspiration

precipitation
soil moisture recharge
soil moisture storage
soil moisture utilization
surplus
transpiration
wilting point
withdrawal

Objectives

After completion of this lab, you should be able to:

1. *Identify* and *define* the key water-balance components as arranged in a water-balance equation.
2. *Compare* and *contrast* maps of precipitation and potential evapotranspiration in North America.
3. *Plot* water-balance components for Kingsport, TN, and Sacramento, CA, and *graph* the relationship between PRECIP and POTET.
4. *Explain* the dynamics of soil moisture storage changes—recharge and utilization.

Materials/Sources Needed

calculator
pencil
color pencils

Lab Exercise and Activities

✳ SECTION 1

Water Balance Components

A **soil-water budget** can be established for any area of Earth's surface—a continent, country, region, field, or front yard. Key is measuring the precipitation input and its distribution to satisfy the "demands" of plants, evaporation, and soil moisture storage in the area considered. Such a budget can examine any time frame, from minutes to years.

Think of a soil-water budget as a money budget: precipitation income must be balanced against expenditures of evaporation, transpiration, and runoff. Soil-moisture storage acts as a savings account, accepting deposits and withdrawals of water. Sometimes all expenditure demands are met, and any extra water results in a surplus. At other times, precipitation and soil moisture income are inadequate to meet demands, and a deficit, or water shortage, results.

The water balance describes how the water supply is expended. Think of **precipitation** as "income" and **evapotranspiration** as "expenditure." If income exceeds expenditures, then there is a **surplus** to account for in the budget. If income is not enough to meet demands, then we need to turn to savings (a storage account), if available, to meet these demands. When savings are not available, then we must record **a deficit** of unmet demand.

In the water balance these budgetary components are presented as follows:

- **Precipitation** = income (supply)
- **Potential evapotranspiration** =

    expenditure (demand)
- **Deficit** = unmet expenditures (shortages)
- **Surplus** = excess (oversupply)
- **Soil Storage** = savings

To understand the water-balance methodology and "accounting" or "bookkeeping" procedures, we must first understand the terms and concepts in a simple water-balance equation. Please consult *Geosystems*, Chapter 9, or *Elemental Geosystems*, Chapter 7, for the full description of each of these concepts and how they are measured and determined.

The object, as with a money budget, is to account for the ways in which this supply is distributed: actual water taken by evaporation and plant transpiration, extra water that exits in streams and subsurface groundwater, and recharge or utilization of soil-moisture storage.

Remember the object of the water balance is to account for the expenditure of precipitation. The pronounceable acronyms in the water-balance equation should streamline having to repeat the full terms each time you say them.

Water Balance Equation:

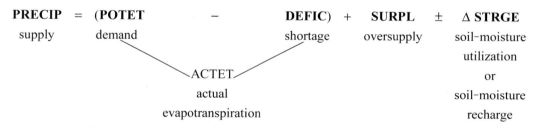

- <u>PRECIP</u> (**precipitation**) is rain, sleet, snow, and hail—*the moisture supply*.
- <u>POTET</u> (**potential evapotranspiration**) is the amount of moisture that would evaporate and transpire through plants if the moisture were available; the amount that would become output under optimum moisture conditions—*the moisture demand*.
- <u>DEFIC</u> (**deficit**) is the amount of unsatisfied POTET; the amount of demand that is not met either by PRECIP or by soil moisture storage—*the moisture shortage*.
- <u>ACTET</u> (**actual evapotranspiration**) is the actual amount of evaporation and transpiration that occurs; derived from POTET – DEFIC; thus, if all the demand is satisfied, POTET will equal ACTET—*the actual satisfied demand*.

- SURPL (**surplus**) is the amount of moisture that exceeds POTET, when soil moisture storage is at field capacity (full)—*the moisture oversupply*.

- ±Δ STRGE (**soil moisture storage change**) is the use (decrease) or recharge (increase) of soil moisture, snow pack, or lake and surface storage or detention of water—*the moisture savings*.

Key to the water balance is determining the amount of water that would evaporate and transpire if it were available (POTET). Please compare the PRECIP (supply) map in Figure 10.1a to the POTET (demand) map in Figure 10.1b for the United States and Canada. The relationship between PRECIP supplies and POTET demands determines the remaining components of the water-balance equation—water resources.

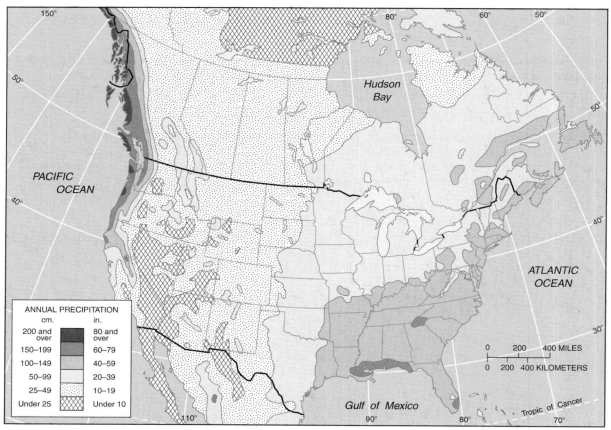

(a) Precipitation

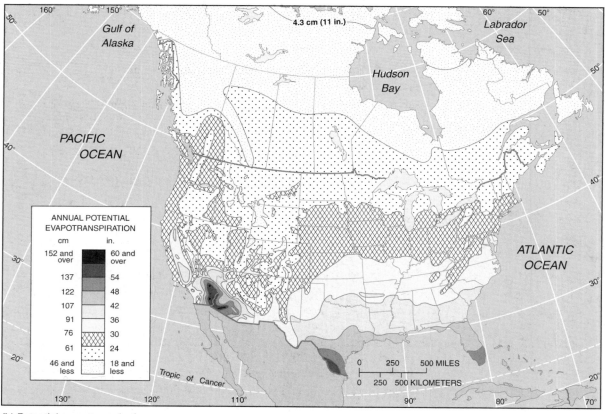

(b) Potential evapotranspiration

Figure 10.1a and b

Potential evapotranspiration and precipitation for the United States and Canada.

Complete the following questions about water-balance and the PRECIP and POTET maps (Figures 10.1a and 10.1b) for North America:

1. Can you identify from the two maps regions where PRECIP <u>supply</u> (10.1a) is <u>higher</u> than POTET <u>demand</u> (10.1b)? Compare the two maps and describe these regions. _____

   _____

   _____

   _____

2. Can you identify from the two maps regions where POTET <u>demand</u> is <u>higher</u> than PRECIP <u>supply</u>? Compare the two maps and describe these regions. _____

   _____

   _____

   _____

3. Based on these maps, why does 95% of the irrigated agriculture in the United States and Canada occur west of the 95th meridian (central Kansas, western Manitoba)? _____

   _____

   _____

4. Where you go to college, is the natural water demand usually met by the natural precipitation supply? Or, does your region experience a natural shortage? Are there some months of surplus and some months of deficit in the annual pattern? _____

   _____

   _____

5. What kinds of adaptations are made in your area to overcome the natural shortage? For instance, do people have to install sprinkler systems for lawns, or is natural precipitation adequate all year long? _____

   _____

   _____

✳ SECTION 2

Water Balance Supply and Demand for Kingsport, Tennessee

Table 10.1 presents the long-term average water supply and demand conditions for the city of Kingsport in the extreme northeast corner of Tennessee. Kingsport is located at 36.5° N, 82.5° W, with an elevation of 390 m (1280 ft), and has maintained weather records for more than half a century.

The monthly values in Table 10.1 assume that PRECIP and POTET are evenly distributed throughout each month, smoothing the actual daily and hourly conditions. Kingsport experiences a humid subtropical climate with adequate moisture throughout the year.

Let's first compare PRECIP with POTET by month to determine whether there is a net supply (+) or net demand (−) for water. This is the basic bookkeeping procedure. Calculate the positive or negative relationship between PRECIP and POTET. Record this plus or minus value in the spaces provided—note January and June are already figured.

Table 10.1
PRECIP and POTET for Kingsport

	Jan	Feb	Mar	Apr	May	Jun	Jul	Aug	Sep	Oct	Nov	Dec
PRECIP	97	99	97	84	104	97	132	112	66	66	66	99
POTET	7	8	24	57	97	132	150	133	99	55	12	7
Net supply (+)	+90	+91	+73	+27	+7	—	—	—	—	+11	+54	+92
Or demand (−)	—	—	—	—	—	−35	−18	−21	−33	—	—	

(All quantities in millimeters)

Precip *Potet*
110 *↑ 40*

Questions about Kingsport's water-balance accounting.

↓ NetSupply
70
Initial Storage: 40mm
EOM : 100mm

1. When does Kingsport experience a net supply of water? List the months.

Jan, Feb, March, Apr, May, Oct, Nov, Dec

2. What occurs during the warm months from June through September?

Less water, less energy.

3. What is the total annual PRECIP for Kingsport? ____1119____ mm; __28,422.6__ in.

4. What is the total annual POTET for Kingsport? ____781____ mm; __19,837.4__ in.
[25.4 mm = 1 in.]

ACTET − (water vaper)
              (energy vaper)

Water Budget Calculations for Kingsport, Tennessee

Soil-moisture storage is a "savings account" of water that can receive deposits and allow withdrawals as conditions change in the water balance. Soil-moisture storage (ΔSTRGE) refers to the amount of water that is stored in the soil and is accessible to plant roots. Soil is said to be at the **wilting point** when all that is left in the soil is unextractable water; the plants wilt and eventually die after a prolonged period of such moisture stress.

The soil moisture that is generally accessible to plant roots is **capillary water**, held in the soil by surface tension and hydrogen-bonding between the water and the soil. Almost all capillary water is **available water** in soil moisture storage and is removable for POTET demands through the action of plant roots and surface evaporation. After water drains from the larger pore spaces, the available water remaining for plants is termed **field capacity**, or storage capacity. This water is held in the soil by hydrogen bonding against the pull of gravity. Field capacity is specific to each soil type and is an amount that can be determined by soil surveys. If you have a garden or house plant, you no doubt are aware of how much effort is needed to maximize soils for good water retention.

Assuming a soil moisture storage capacity of 100 mm (4.0 in.) for Kingsport, Tennessee, typical of shallow-rooted plants, the months of net demand for moisture are satisfied through **soil-moisture utilization**. Various plant types send roots to different depths and therefore are exposed to varying amounts of soil moisture.

For example, shallow-rooted crops such as spinach, beans, and carrots send roots down about 65 cm (25 in.) in a silt loam, whereas deep-rooted crops such as alfalfa and shrubs exceed a depth of 125 cm (50 in.) in such a soil.

For this exercise we assume that soil moisture utilization occurs at 100%, that is, if there is a net water demand, the plants will be able to extract moisture as needed. Actually, in nature as the available soil water is reduced by soil-moisture utilization, the plants must exert greater effort to extract the same amount of moisture. As a result, even though a small amount of water may remain in the soil, plants may be unable to exert enough pressure to utilize it. The unsatisfied demand resulting from this situation is calculated as a *deficit*. Avoiding such deficit inefficiencies and reduction in plant growth are the goals of a proper irrigation program, for the harder plants must work to get water, the less their yield and growth will be.

Likewise, relative to soil moisture recharge we assume a 100% rate if the soil moisture storage is less than field capacity, then excess moisture beyond POTET demand will go to **soil-moisture recharge**. We assume in this exercise a soil moisture recharge rate as 100% efficient as long as the soil is below field capacity and above a temperature of $-1°C$ (30.2°F). Under real conditions we know that infiltration actually proceeds rapidly in the first minutes of a storm, slowing as the upper layers of soil become saturated even though the soil below is still dry.

Questions and analysis about the water balance of Kingsport, Tennessee.

1. With these assumptions about the operation of the soil-moisture resource, please take your PRECIP—POTET calculations from Table 10.1 and carry them over to Table 10.2 on the third line of the table. Complete the remainder of the water-balance table, recalling that storage capacity is 100 mm. Data are smoothed over the entire month.

Table 10.2
Kingsport Water Balance

Net supply →
Demand →
change in → storage.

	Jan	Feb	Mar	Apr	May	Jun	Jul	Aug	Sep	Oct	Nov	Dec	Total
PRECIP	97	99	97	84	104	97	132	112	66	66	66	99	1119
POTET	7	8	24	57	97	132	150	133	99	55	12	7	781
PRECIP—POTET	+90	91	+73	+27	+7	−35	−18	−21	−33	+11	+54	+92	—
STRGE end of month	100	100	100	100	100	65	47	26	0 diff	11	65	100	—
ΔSTRGE	0	0	0	0	0	−35	−18	−21	−26	11	54	+35	←
ACTET	7	8	24	57	97	132	150	133	92	55	12	7	—
DEFIC	0	0	0	0	0	0	0	0	7	0	0	0	—
SURPL	90	91	73	27	7	0	0	0	0	0	0	57	—

(All quantities in millimeters)

2. Soil moisture remains at field capacity (full) through which month? _Jan, Feb, Mar, Apr, May, Dec_

3. How much surplus is accumulated through these first five months? _90+91+73+27+7= 288_

4. What is the net demand for water in June? _−35m_

   After you satisfy this demand through soil moisture utilization, what is the remaining water in soil moisture

   at the end of June, to begin the month of July? _Same_

5. Calculate the actual evapotranspiration for each month of the year for Kingsport and note this on the table. By subtracting DEFIC from POTET, you determine the *actual evapotranspiration*, or ACTET, that takes place for each month. Under ideal moisture conditions POTET and ACTET are about the same, so that plants do not experience a water shortage. Prolonged deficits could lead to drought conditions.

   You will note, that given the 100% efficiency assumptions for soil-moisture utilization used in this exercise, ACTET equals POTET each month. In reality there actually would be a slight inefficiency of soil-moisture utilization from June through September, with ACTET progressively less than the POTET producing a four-month deficit of 16 mm.

6. According to your calculations, do the soils of Kingsport return to field capacity (full storage) by the end of

   the year? _Yes_ _____ Are any surpluses generated in December?

   The amount? _57m_

7. What is the total ACTET for the year? _874_

8. What is the total DEFIC for the year? _7_

9. What is the total SURPL for the year? _345_

10. During how many months at Kingsport does POTET exceed PRECIP? ___Two Months___

11. According to a physical geography textbook, and lab discussion, what precipitation mechanisms produce Kingsport's pattern of summer maximum precipitation and moisture throughout the year? What factors produce such consistent precipitation throughout the year?

Winter: _____

_____

Summer: _____

_____

✳ SECTION 4

Water Budget Calculations for Sacramento, California

For comparison let's work with a city that experiences seasonal deficits in its annual water balance. Sacramento, California, (38.58° N, 121.35° W, at 11 m elevation) has a Mediterranean dry, warm summer climate. The data for Sacramento are set in Table 10.3 similar to Kingsport. Please assume the same soil-moisture storage capacity of 100 mm (4.0 in.) for Sacramento, typical of shallow-rooted plants. The months of net demand for moisture are satisfied through *soil-moisture utilization*, as long as the soil moisture is available. Sacramento does experience a wilting point each year.

Table 10.3
Water Balance for Sacramento, California

	Jan	Feb	Mar	Apr	May	Jun	Jul	Aug	Sep	Oct	Nov	Dec	Total
PRECIP	107	74	56	36	10	3	3	3	8	23	58	76	455
POTET	15	23	40	60	85	115	140	127	99	66	30	15	815
PRECIP—POTET													—
STRGE	100					0						89	—
ΔSTRGE													
ACTET													
DEFIC													
SURPL													

(All quantities in millimeters)

Questions and analysis about the water balance of Sacramento, California:

1. Complete the bookkeeping procedure for Sacramento using the monthly and annual average values given in the table. Calculate PRECIP—POTET, ΔSTRGE ACTET, DEFIC, and SURPL amounts for each month as you did for Kingsport.

2. For Sacramento, how many months does POTET exceed PRECIP? _____

3. According to a physical geography textbook, and lab discussion, what precipitation mechanisms produce Sacramento's seasonally variable summer-drought, winter-moist precipitation pattern?

Winter: _____

_____

Summer: _____

_____

✳ SECTION 5

Water Balance Graphs

A useful way to visualize the water balance for a city is to graph the data. The following activity will allow you to graph and then compare the water balances for Kingsport and Sacramento.

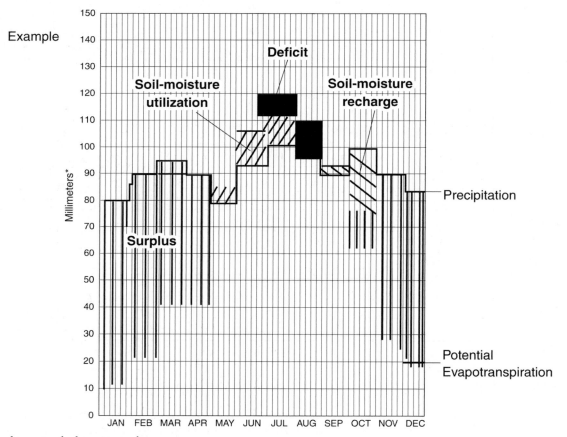

Sample water balance graph

1. Take the PRECIP and POTET data presented in ✳ SECTION 3 and 4 for Kingsport and Sacramento and prepare a water balance graph for each station. Since the data are given as monthly totals, assume an even distribution of supply and demand over each month—a stepwise graph. Prepare the graphs as line graphs by month. Use a *blue line* for PRECIP and a *red line* for POTET. (See the graph above for an example of a stepwise graph.)

(a)

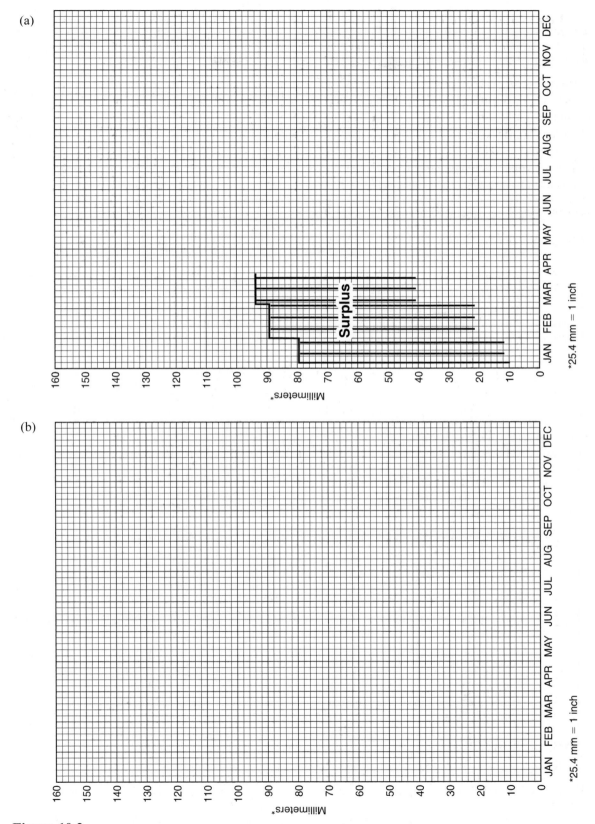

Surplus

*25.4 mm = 1 inch

(b)

*25.4 mm = 1 inch

Figure 10.2
(a) Water balance graph for Kingsport, Tennessee. (b) Water balance graph for Sacramento, California.

2. Identify with shading the areas *between* the PRECIP and POTET line-graph plots in #1 above, that represent various aspects of each water balance (refer to the example at the beginning of ✳ SECTION 5 to help you). For the four relations possible between moisture supply and demand, utilize the following key colors for shading the appropriate portions of your graphs:

- **Surplus**: blue shading (PRECIP exceeds POTET)
- **Soil moisture utilization**: brown shading (POTET exceeds PRECIP with soil moisture available to meet some of the demand)
- **Deficit**: orange shading (POTET exceeds PRECIP with inadequate soil moisture available, resulting in moisture shortages)
- **Soil moisture recharge**: green shading (PRECIP exceeds POTET and excess water entering the soil storage until it reaches field capacity)

Questions about these water balance graphs:

3. Describe the pattern (months and seasons) of water surpluses for Sacramento:

_____

_____

4. Describe the pattern of water deficits (months and seasons) for Sacramento:

_____

_____

5. Water resources, the "water crop," are harvested from water surplus. If you were a water resource manager for the Sacramento region, what strategies would you recommend to meet agricultural and urban water demands? (Discuss this among others in your lab before you begin writing. Note that there is a mountain range east of the Sacramento region that accumulates a snow pack in winter; make this part of your consideration.)

_____

_____

_____

_____

6. If you were a lawn-sprinkler sales person, in which city would you choose to establish your business? Explain your choice.

_____

_____

_____

Name: _____

Date: _____   Score/Grade: _____

Lab Exercise 11

Global Climate Systems

Earth experiences an almost infinite variety of *weather*—conditions of the atmosphere at any given time and place. But if we consider the weather over many years, including its variability and extremes, a pattern emerges that constitutes **climate**. Think of climate patterns as dynamic rather than static. Climate is more than a consideration of simple averages of temperature and precipitation.

Climatology, the study of climate, is the analysis of long-term weather patterns, including extreme weather events, over time and space to find areas of similar weather statistics, identified as **climatic regions**. Observed patterns grouped into regions are at the core of climate classification.

In this lab exercise, we examine various patterns of temperature and precipitation that operate as a basis for climate **classification**. We work with a model of global climates, the classification and plotting of actual climate data, analysis of temperature and precipitation patterns. Lab Exercise 11 features six sections (one optional).

Key Terms and Concepts

classification
climate
climatic regions
climatology

climographs
empirical classification
genetic classification

Objectives

After completion of this lab, you should be able to:

1. *Identify* patterns of temperature, precipitation, and other weather elements that contribute to the climate of an area.
2. *Analyze* key climate data to *determine* the empirical (Köppen) classification and genetic (causative) factors.
3. *Plot* data (°C and cm) on climographs for analysis.
4. *Analyze* patterns of climate change and their effects.

Materials/Sources Needed

color pencils
calculator
world atlas
physical geography text

Lab Exercise and Activities

✳ SECTION 1

Climatology and Climate Classification

Climatology, the study of climate, involves analysis of the patterns in time or space created by various physical factors in the environment. One type of climatic analysis involves discerning areas of similar weather statistics and grouping these into climatic regions that contain characteristic weather patterns. Climate classifications are an effort to formalize these patterns and determine their implications to humans. Simply comparing the two principal climatic components—temperature and precipitation—reveals important relationships indicative of the distribution of climate types (Figure 11.1).

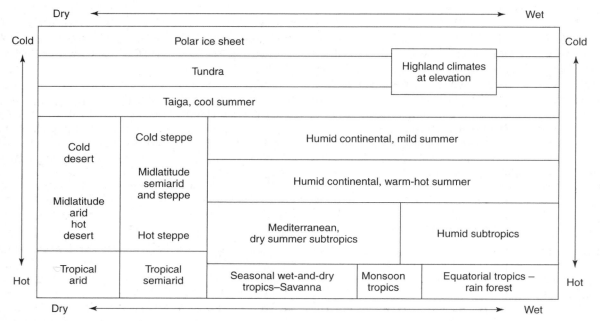

Figure 11.1
Temperature and precipitation schematic showing climatic relationships

1. Mark your present location with an **X** on Figure 11.1. Place the name of this location next to mark. Next, mark the location (approximately) that is most characteristic of your place of birth with an **O** and record the name of that place next to the mark.

Classification is the process of grouping data or phenomena in related classes. A classification based on causative factors—for example, the interaction of air masses—is called a **genetic classification**. An **empirical classification** is based on statistics or other data used to determine general categories. Climate classifications based on temperature and precipitation data are empirical classifications. In *Applied Physical Geography,* we use Köppen's empirical system to classify climatic regions, as well as a genetic analysis to help you appreciate the relationships between a region's location and its climate.

The principal elements of climate are insolation, temperature, pressure, air masses, and precipitation. We review them briefly here. Insolation is the energy input for the climate system, but it varies widely over Earth's surface by latitude (see Figure 4.3). The pattern of world temperatures and their annual ranges is shown in Figure 6.4a and Figure 6.4b. Temperature variations result from a coupling of dynamic forces in the atmosphere to Earth's pattern of atmospheric pressure and resulting global wind systems (see Figures 7.2 and 7.3). Important, too, are the location and physical characteristics of air masses, those vast bodies of homogeneous air that form over oceanic and continental source regions.

Moisture is the remaining input to climate. The hydrologic cycle transfers moisture, with its tremendous latent heat energy, through Earth's climate system. Figure 10.3 in *Geosystems,* eighth edition, and Figure 6.2 in *Elemental Geosystems,* sixth edition, show the worldwide distribution of precipitation, our moisture supply. Moisture patterns are important, for they are key climate control factors. Average temperatures and day length help us approximate POTET (potential evapotranspiration), a measure of natural moisture demand. Most of Earth's desert regions, areas of permanent water deficit, are in lands dominated by subtropical high pressure cells, with bordering lands grading to grasslands and to forests as precipitation increases. The most consistently wet climates on Earth straddle the equator in the Amazon region of South America, the Congo region of Africa, and Indonesia and Southeast Asia, all of which are influenced by equatorial low pressure and the intertropical conver-

gence zone or ITCZ. Simply relating the two principal climatic components—temperature and precipitation—reveals general climate types.

The basis of any empirical classification system is the choice of criteria used to draw lines between categories. The Köppen classification system, widely used for its ease of comprehension, was designed by Wladimir Köppen (1846–1940), a German climatologist and botanist. Köppen chose mean monthly temperatures, mean monthly precipitation, and total annual precipitation to devise his spatial categories and boundaries. But we must remember that boundaries really are gray areas; they are transition zones of gradual change. For our purposes of general understanding, a modified Köppen-Geiger classification is useful.

The Köppen system uses capital letters (A, B, C, D, E, plus H) to designate climatic categories from the equator to the poles. Five of the climate classifications are based upon thermal criteria and one on moisture conditions as well.

A—Tropical (equatorial regions)
C—Mesothermal (Mediterranean, humid subtropical, marine west coast regions)
D—Microthermal (humid continental, subarctic regions)
E—Polar (polar regions)
H—Highland (compared to lowlands at the same latitude, highlands have lower temperatures—recall the normal lapse rate—and more efficient precipitation due to lower POTET demand, justifying separation of highlands from surrounding climates)
B—Dry, the deserts and steppes (only category based on moisture as well)

In addition to these primary types noted with capital letters, lowercase letters are used to signify certain temperature and moisture characteristics. Inside the back cover of this laboratory manual is a world map illustrating climates classified according to the Köppen system, with the addition of the main climatic influences. Table 11.1 presents a climate classification key and summary that specifies both *thermal* (A, C, D, and E) and *moisture* (B) classifications. Use this table for classifying each of the five stations presented in Lab Exercise ✳ SECTION 2 and two stations in ✳ SECTION 4. Specific formulas to calculate the *B* dry climates follow in this Lab Exercise in ✳ SECTION 4.

For example, in a *tropical rain forest Af* climate, the *A* tells us that the average coolest month is above 18°C (64.4°F, average for the month), and the *f* indicates that the weather is constantly wet (German feucht, for "moist"), with the driest month receiving at least 6 cm (2.4 in.) of precipitation. As you can see in

the map of A climates, the *tropical rain forest Af* climate straddles the equator.

For another example, in a *Dfa* climate, the *D* means that the average warmest month is above 10°C (50°F), with at least one month falling below 0°C (32°F); the *f* says that at least 3 cm (1.2 in.) of precipitation falls during every month; and the *a* indicates a warmest summer month averaging above 22°C (71.6°F). Thus, a *Dfa* climate is a humid-continental, hot-summer climate in the microthermal D category.

Try not to get lost in the alphabet soup of this system, for it is meant to help you understand complex climates through a simplified set of symbols. To simplify the detailed criteria, a box is placed at the end of each climate category that presents Köppen's guidelines. To help keep it straight, always say the climate's descriptive name followed by its symbol: for example, *Mediterranean dry-summer Csa, tundra ET,* or *cold midlatitude steppe BSk.*

Determining primary climate classification

E An **E climate** if warmest month is less than 10° C (50° F). If not an E then move on to B climates.

B If precipitation supply is less than average potential evapotranspiration demand climate may be a **B climate**—go to Lab ★**STEP 5** to consult formulas. If climate is more moist than B criteria then move on to A climates.

A An **A climate** if the average coolest month is warmer than 18° C (64.4° F). If not an A then move on to C climates.

C A **C climate** if average coldest month is below 18° C (64.4° F) but warmer than 0° C (32° F). If not an E, B, A or C then climate is a D classification.

D A **D climate** if average coldest month is below 0° C (32° F) and warmest month is above 10° C (50° F).

E Polar Climates

Warmest month below 10° C (50° F); always cold; ice climates.

EF - Ice cap:
Warmest month between 0° C (32° F); PRECIP exceeds very small POTET demand; the polar regions.

ET - Tundra:
Warmest month between 0-10° C (32-50° F); PRECIP exceeds small POTET demand; snow cover 8-10 months.

EM - Polar marine:
All months above −7° C (20° F), warmest month above 0° C; annual temperature range <17° C (30° F).

B Dry Arid and Semiarid Climates

POTET exceeds PRECIP in all B climates. Subdivisions based on PRECIP timing and amount and mean annual temperature. Boundaries determined by formulas and graphs.

Earth's arid climates.	**Earth's semiarid climates.**
BWh - Hot low-latitude desert	BSh - Hot low-latitude steppe
BWk - Cold midlatitude desert	BSk - Cold midlatitude steppe
BW = PRECIP less than 1/2 POTET.	BS = PRECIP MORE THAN 1/2 POTET but not equal to it.
h = Mean annual temperature >18° C (64.4° F).	h = Mean annual temperature >18° C.
k = Mean annual temperature < 18° C.	k = Mean annual temperature < 18° C.

A Tropical Climates

Consistently warm with all months averaging above 18° C (64.4° F); annual PRECIP exceeds POTET.

Am - Tropical monsoon:
m = A marked short dry season with 1 or more months receiving less than 6 cm (2.4 in.) PRECIP; an otherwise excessively wet rainy season. ITCZ 6–12 months dominant.

Af - Tropical rain forest:
f = All months receive PRECIP in excess of 6 cm (2.4 in.).

Aw - Tropical savanna:
w = Summer wet season, winter dry season; ITCZ dominant 6 months or less, winter water-balance deficits.

C Mesothermal Climates

Warmest month above 10° C (50° F); coldest month above 0° C (32° F) but below 18° C (64.4° F); seasonal climates.

Cfb, Cfc - Marine west coast: Mild-to-cool summer.
f = Receives year-round PRECIP
b = Warmest month below 22° C (71.6° F) with 4 months above 10° C.
c = 1 to 3 months above 10° C.

Cfa, Cwa - Humid subtropical:
a = Hot summer: warmest month above 22° C (71.6° F).
f = Year-round PRECIP.
w = Winter drought, summer wettest month 10 times more PRECIP than driest winter month.

Csa, Csb - Mediterranean summer dry:
s = Pronounced summer drought with 70% of PRECIP in winter.
a = Hot summer with warmest month above 22° C (71.6° F).
b = Mild summer; warmest month below 22° C.

D Microthermal Climates

Warmest month above 10° C (50° F); coldest month below 0° C (32° F); cool temperate-to-cold conditions; snow climates. In Southern Hemisphere, only in highland climates.

Dfb, Dwb - Humid continental:
b = Mild summer; warmest month below 22° C (71.6° F).
f = Year-round PRECIP.
w = Winter drought.

Dfa, Dwa - Humid continental:
a = Hot summer; warmest month above 22° C (71.6° F).
f = Year-round PRECIP.
w = Winter drought.

Dfc, Dwc, Dwd - Subarctic:
Cool summers, cold winters.
f = Year-round PRECIP.
w = Winter drought.
c = 1 to 4 months above 10° C.
d = Coldest month below −38° C (−36.4° F), in Siberia only.

Table 11.1
Climate Classification Key

Climographs for Five Stations

Temperature patterns are influenced by several principal controls: latitudinal position, altitude, cloud cover, land-water heating differences, ocean currents and sea-surface temperatures, and general surface conditions. Review your work in Lab Exercise 6 with stations in Bolivia and Canada where you identified the effects of these controls.

Use the five climographs (Figures 11.2a through 11.2e) and sets of data that follow to plot mean monthly temperature (*solid-line graph*), mean monthly precipitation (*a bar graph*), and potential evapotranspiration (*a dotted line*) data for the following stations. Use Table 11.1 to determine each station's climatic classification and then determine the city that identifies each station. Analyze the distribution of temperature and precipitation during the year and, using your atlas and the map inside the back cover of this laboratory manual, find and record the names of the five other cities listed below.

Selected Global Climate Stations

✳**SECTION 2**—Tropical, mesothermal, and microthermal climographs, six stations Salvador (Bahia), Brazil: pop. 2,676,000; elev. 9 m (30 ft), completed as an example.

Figures 11.2a through e—five stations to graph and identify (listed in alphabetical order)

Edinburgh, Scotland: pop. 478,000; elev. 134 m (439 ft), 56°N 3.1°W

Montreal, Quebéc, Canada: pop. 1,621,000; elev. 57 m (187 ft), 45.5°N 73.5°W

New Orleans, Louisiana: pop. 337,000; elev. 3 m (9 ft), 30°N 90°W

Oymyakon, Siberia, Russia: pop. 521; elev. 726 m (2382 ft), 63°N 143°E

Sacramento, California: pop. 486,000; elev. 11 m (36 ft), 38.5°N 121.3°W

✳**SECTION 3**—Climate analysis

✳**SECTION 4**—Desert climate climographs, two stations

Figures 11.4a and b—two stations to graph and identify (listed in no particular order)

Reno, Nevada: pop. 221,000; elev. 1341 m (4400 ft), 39.5°N 119.5°W

Walgett, New South Wales, Australia: pop. 2000; elev. 133 m (436 ft), 30°S 148°E

✳**SECTION 5**—Your climate region (optional station exercise)

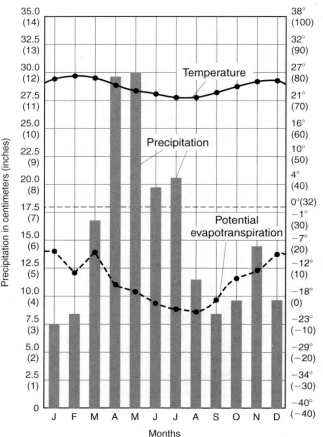

Figure 11.2

Example climograph:

Station #1: Salvador (Bahia), Brazil

Total annual rainfall: 183.6 cm (72.3 in.)

Average annual temperature: 24.8°C (76.6°F)

Annual temperature range: 3.4 C° (6.1°F)

Distribution of temperature during the year:

Consistently warm; diurnal

Range: greater than daily range

Distribution of precipitation during the year:

Large total, maximum

in April and May

Principal atmospheric lifting mechanism(s):

Convergent lifting during wetter season

with convectional lifting the rest of the year

Distribution of potential evapotranspiration during
the year: Consistent, never below 8.6 cm (3.4 in.)

	Jan	Feb	Mar	Apr	May	Jun	Jul	Aug	Sep	Oct	Nov	Dec	Annual
Temperature °C	26.0	26.3	26.3	25.8	24.8	23.8	23.0	22.9	23.6	24.5	25.1	25.6	24.8
(°F)	(78.8)	(79.3)	(79.3)	(78.4)	(76.6)	(74.8)	(73.4)	(73.2)	(74.5)	(76.1)	(77.2)	(78.1)	(76.6)
PRECIP cm	7.4	7.9	16.3	29.0	29.7	19.6	20.6	11.2	8.4	9.4	14.2	9.9	183.6
(in.)	(2.9)	(3.1)	(6.4)	(11.4)	(11.7)	(7.7)	(8.1)	(4.4)	(3.3)	(3.7)	(5.6)	(3.9)	(72.3)
POTET cm	13.7	12.4	13.5	11.9	10.7	9.1	8.6	8.6	9.4	11.2	11.9	13.2	134.4
(in.)	(5.4)	(4.9)	(5.3)	(4.7)	(4.2)	(3.6)	(3.4)	(3.4)	(3.7)	(4.4)	(4.7)	(5.2)	(52.9)

What are the main climatic influences for this station (air pressure, air mass sources, degree of continentality,

temperature of ocean currents)? _____

_____

Köppen climate classification symbol: <u>Af</u>; name: _____Tropical rain forest_____

explanation for this determination: All months above 18°C (64.4°F) and all months receive PRECIP over 6 cm

(2.4 in.).

Representative biome (terrestrial ecosystem) characteristic of region: ETR, equatorial and tropical rain forest;

characteristic vegetation: _____evergreen rain forest_____

Figure 11.2a

Station #2: _____

Total annual rainfall: _____

Average annual temperature: _____

Annual temperature range: _____

Distribution of temperature during the year:

_____

_____

Distribution of precipitation during the year:

_____

_____

Principal atmospheric lifting mechanism(s):

_____

_____

Distribution of potential evapotranspiration during

the year: _____

_____

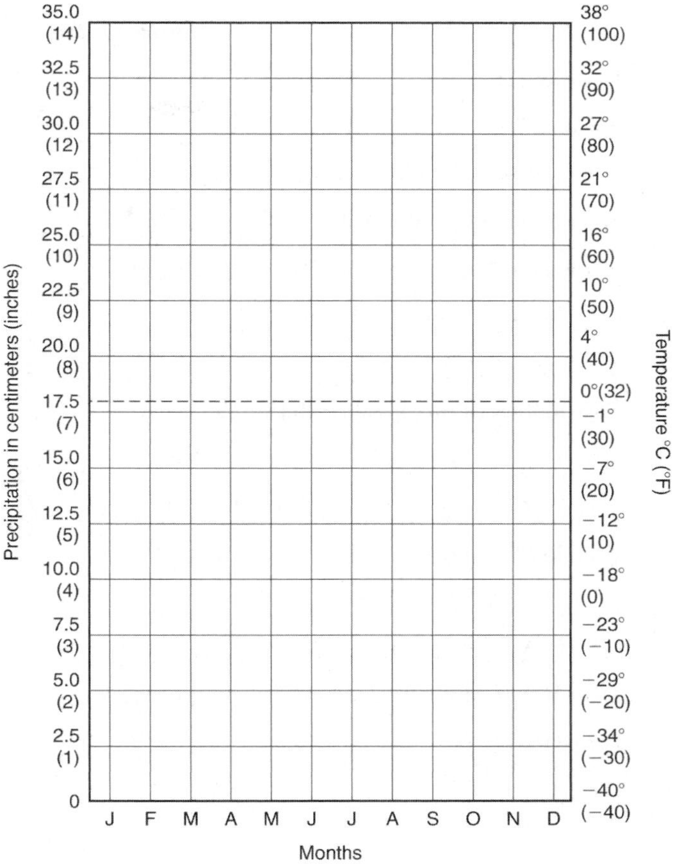

	Jan	Feb	Mar	Apr	May	Jun	Jul	Aug	Sep	Oct	Nov	Dec	Annual
Temperature °C	−10.0	−9.4	−3.3	5.6	13.3	18.3	22.1	19.4	15.0	8.3	0.6	−6.7	6.0
(°F)	(14.0)	(15.1)	(26.1)	(42.1)	(55.9)	(64.9)	(71.7)	(66.9)	(59.0)	(46.9)	(33.1)	(19.9)	(42.8)
PRECIP cm	9.6	7.7	8.8	6.6	8.0	8.7	9.5	8.8	9.3	8.7	9.0	9.1	103.8
(in.)	(3.8)	(3.0)	(3.5)	(2.6)	(3.1)	(3.4)	(3.7)	(3.5)	(3.7)	(3.4)	(3.5)	(3.6)	(40.8)
POTET cm	0	0	0	2.7	8.1	11.9	13.9	12.1	7.7	3.7	0.2	0	60.3
(in.)	(0)	(0)	(0)	(1.1)	(3.2)	(4.7)	(5.5)	(4.8)	(3.0)	(1.5)	(0.1)	(0)	(23.7)

What are the main climatic influences for this station (air pressure, air mass sources, degree of continentality, temperature of ocean currents)? _____

_____

_____

Köppen climate classification symbol: _____ ; name: _____

_____ ; explanation for this determination: _____

_____

_____

Representative biome (terrestrial ecosystem) characteristic of region: _____

_____ ; characteristic vegetation: _____

Figure 11.2b

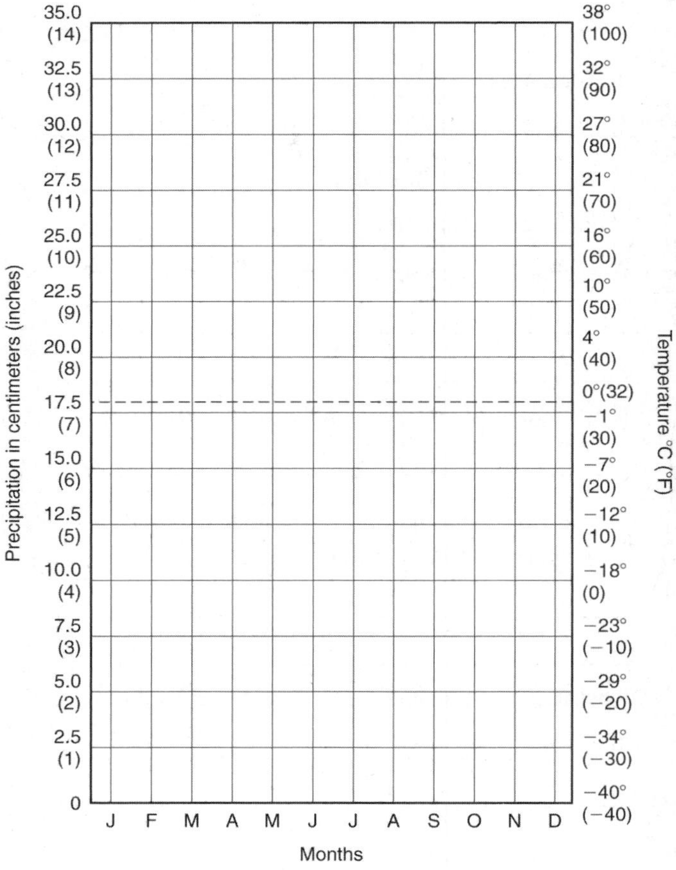

Station #3: _____

Total annual rainfall: _____

Average annual temperature: _____

Annual temperature range: _____

Distribution of temperature during the year:

_____

_____

Distribution of precipitation during the year:

_____

_____

Principal atmospheric lifting mechanism(s):

_____

_____

Distribution of potential evapotranspiration during

the year: _____

_____

	Jan	Feb	Mar	Apr	May	Jun	Jul	Aug	Sep	Oct	Nov	Dec	Annual
Temperature °C	−47.2	−42.9	−34.2	−15.4	1.4	11.6	14.8	10.9	1.6	−16.2	−35.0	−44.0	−16.3
(°F)	−53.0	−45.2	−29.6	(4.3)	(34.5)	(52.9)	(58.6)	(51.6)	(34.9)	(2.8)	−31.0	−47.2	(2.7)
PRECIP cm	0.8	0.5	0.5	0.3	1.0	3.3	4.1	3.8	2.0	1.3	1.0	0.08	19.3
(in.)	(0.3)	(0.2)	(0.2)	(0.1)	(0.4)	(1.3)	(1.6)	(1.5)	(0.8)	(0.5)	(0.4)	(0.3)	(7.6)
POTET cm	0	0	0	0	3.0	13.5	15.7	10.9	2.3	0	0	0	45.5
(in.)	(0)	(0)	(0)	(0)	(1.2)	(5.3)	(6.2)	(4.3)	(0.9)	(0)	(0)	(0)	(17.9)

What are the main climatic influences for this station (air pressure, air mass sources, degree of continentality, temperature of ocean currents)? _____

_____

_____

Köppen climate classification symbol: _____; name: _____

_____; explanation for this determination: _____

_____

_____

Representative biome (terrestrial ecosystem) characteristic of region: _____

_____; characteristic vegetation: _____

Figure 11.2c

Station #4: _____

Total annual rainfall: _____

Average annual temperature: _____

Annual temperature range: _____

Distribution of temperature during the year:

_____

_____

Distribution of precipitation during the year:

_____

_____

Principal atmospheric lifting mechanism(s):

_____

_____

Distribution of potential evapotranspiration during

the year: _____

_____

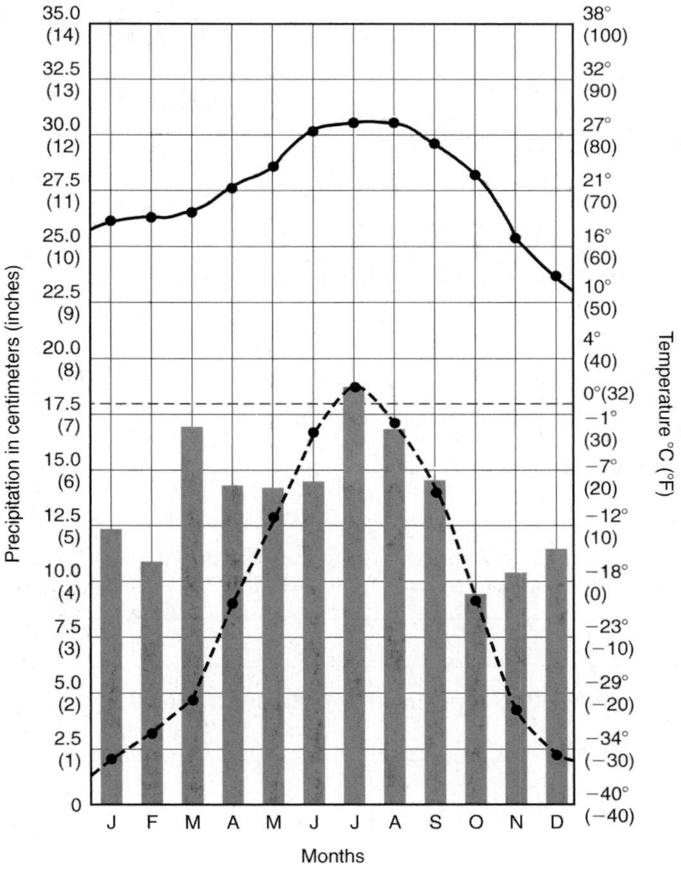

	Jan	Feb	Mar	Apr	May	Jun	Jul	Aug	Sep	Oct	Nov	Dec	Annual
Temperature °C	13.3	14.4	17.2	21.1	24.4	27.8	28.3	28.3	26.7	22.8	16.7	13.9	21.1
(°F)	(56.0)	(58.0)	(63.0)	(70.0)	(76.0)	(82.0)	(83.0)	(83.0)	(80.0)	(73.0)	(62.0)	(57.0)	(70.0)
PRECIP cm	12.2	10.7	16.8	13.7	13.7	14.2	18.0	16.3	14.7	9.4	10.2	11.7	161.3
(in.)	(4.8)	(4.2)	(6.6)	(5.4)	(5.4)	(5.6)	(7.1)	(6.4)	(5.8)	(3.7)	(4.0)	(4.6)	(63.5)
POTET cm	2.2	2.6	4.9	8.4	12.7	16.8	18.0	17.1	13.9	8.8	4.0	2.4	111.8
(in.)	(0.9)	(1.0)	(1.9)	(3.3)	(5.0)	(6.6)	(7.1)	(6.7)	(5.5)	(3.5)	(1.6)	(0.9)	(44.0)

What are the main climatic influences for this station (air pressure, air mass sources, degree of continentality,

temperature of ocean currents)? _____

_____

Köppen climate classification symbol: _____ ; name: _____

_____ ; explanation for this determination: _____

_____

_____

Representative biome (terrestrial ecosystem) characteristic of region: _____

_____ ; characteristic vegetation: _____

Figure 11.2d

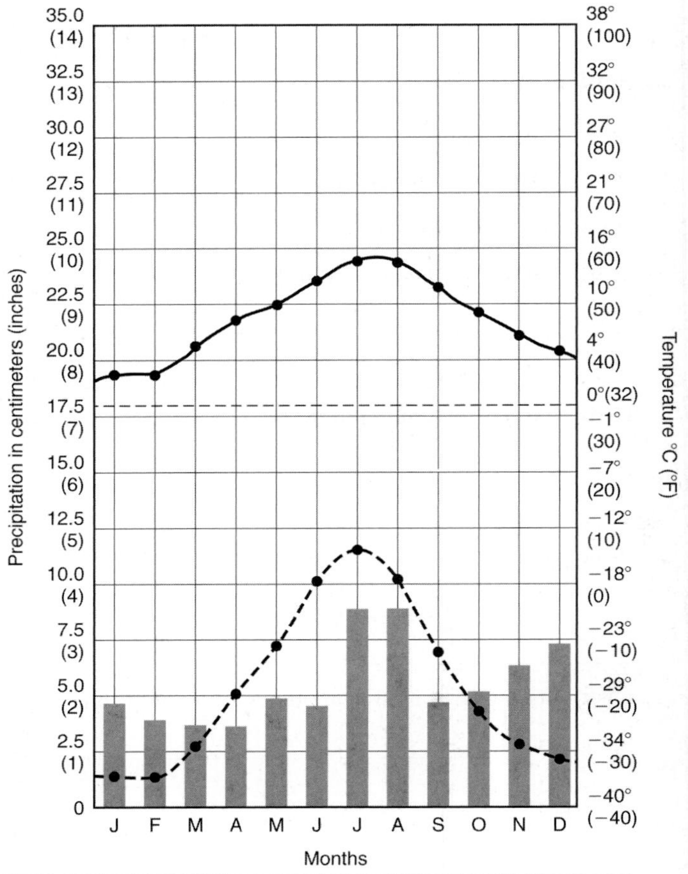

Station #5: _____

Total annual rainfall: _____

Average annual temperature: _____

Annual temperature range: _____

Distribution of temperature during the year:

_____

_____

Distribution of precipitation during the year:

_____

_____

Principal atmospheric lifting mechanism(s):

_____

_____

Distribution of potential evapotranspiration during

the year: _____

	Jan	Feb	Mar	Apr	May	Jun	Jul	Aug	Sep	Oct	Nov	Dec	Annual
Temperature °C	3.0	3.0	5.0	7.6	10.1	12.7	14.7	14.3	12.5	9.7	6.5	4.8	8.7
(°F)	(37.4)	(37.4)	(41.0)	(45.7)	(50.2)	(54.9)	(58.5)	(57.7)	(54.5)	(49.5)	(43.7)	(40.6)	(47.7)
PRECIP cm	4.8	3.6	3.3	3.3	4.8	4.6	8.9	9.1	4.8	5.1	6.1	7.4	65.8
(in.)	(1.9)	(1.4)	(1.3)	(1.3)	(1.9)	(1.8)	(3.5)	(3.6)	(1.9)	(2.0)	(2.4)	(2.9)	(25.9)
POTET cm	1.3	1.3	2.8	4.8	7.6	10.2	11.7	10.2	7.4	4.8	2.8	1.8	64.8
(in.)	(0.5)	(0.5)	(1.1)	(1.9)	(3.0)	(4.0)	(4.6)	(4.0)	(2.9)	(1.9)	(1.1)	(0.7)	(25.5)

What are the main climatic influences for this station (air pressure, air mass sources, degree of continentality, temperature of ocean currents)? _____

_____

Köppen climate classification symbol: _____ ; name: _____

_____ ; explanation for this determination: _____

_____

_____

Representative biome (terrestrial ecosystem) characteristic of region: _____

_____ ; characteristic vegetation:

_____

Figure 11.2e

Station #6: _____

Total annual rainfall: _____

Average annual temperature: _____

Annual temperature range: _____

Distribution of temperature during the year:

_____

_____

Distribution of precipitation during the year:

_____

_____

Principal atmospheric lifting mechanism(s):

_____

_____

Distribution of potential evapotranspiration during

the year: _____

_____

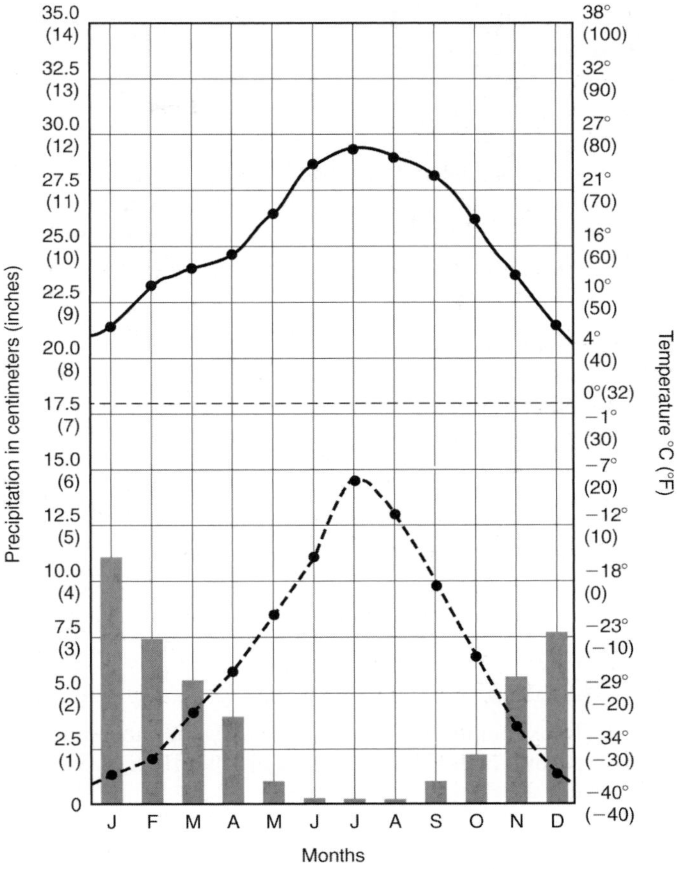

Precipitation in centimeters (inches)

Temperature °C (°F)

Months

	Jan	Feb	Mar	Apr	May	Jun	Jul	Aug	Sep	Oct	Nov	Dec	Annual
Temperature °C	8.4	11.2	12.9	15.6	19.1	22.3	24.8	24.2	22.7	18.5	12.6	8.6	16.8
(°F)	(47.1)	(52.2)	(55.3)	(60.1)	(66.3)	(72.2)	(76.6)	(75.6)	(72.9)	(65.3)	(54.7)	(47.5)	(62.2)
PRECIP cm	10.7	7.4	5.6	3.6	1.0	0.3	0.3	0.3	0.8	2.3	5.8	7.6	45.5
(in.)	(4.2)	(2.9)	(2.2)	(1.4)	(0.4)	(0.1)	(0.1)	(0.1)	(0.3)	(0.9)	(2.3)	(3.0)	(17.9)
POTET cm	1.5	2.3	4.0	6.0	8.5	11.5	14.0	12.7	9.9	6.6	3.0	1.5	81.5
(in.)	(0.6)	(0.9)	(1.6)	(2.4)	(3.3)	(4.5)	(5.5)	(5.0)	(3.9)	(2.6)	(1.2)	(0.6)	(32.1)

What are the main climatic influences for this station (air pressure, air mass sources, degree of continentality, temperature of ocean currents)? _____

_____

_____

Köppen climate classification symbol: _____ ; name: _____

_____ ; explanation for this determination: _____

_____

_____

Representative biome (terrestrial ecosystem) characteristic of region: _____

_____ ; characteristic vegetation: _____

Climate Analysis

Temperature, precipitation, and potential evapotranspiration patterns are the elements used in determining climate classification. Factors influencing these patterns help to more fully understand the distribution of climate regions.

Use *Geosystems* to find background information for some of your responses. Answer the following questions about the completed climographs in Figures 11.2, a through e.

New Orleans—Station # _____

1. What effects do latitudinal position, air mass interactions, and ocean currents have in producing the climate of New Orleans? For instance, why is New Orleans' winter precipitation higher than more northerly cities?

_____

_____

_____

_____

2. In examining POTET (potential evapotranspiration) values throughout the year for New Orleans and comparing them to PRECIP (precipitation) received, do you think there are moisture surpluses or deficits given this data?

_____

_____

_____

Edinburgh—Station # _____

3. Where are the several places in North America that you find this same climatic regime?

_____

_____

_____

4. What controlling factors contribute to the cool summers in Edinburgh? _____

_____

_____

_____

Sacramento—Station # _____

5. What factors produce the natural summer drought (high POTET and low PRECIP) experienced in Sacramento? (Consider air masses, pressure systems, and ocean currents.)

_____

_____

_____

6. How is it possible to have extensive agricultural activity in this region given the summer climate conditions?

_____

_____

_____

_____

Montreal—Station # _____

7. Compare the annual range of temperature (between January and July) in Montreal with your analysis of New Orleans. Why the greater range value for Montreal? _____

_____

_____

_____

8. What contributing factors (mechanisms and conditions) produce the even distribution of precipitation throughout the year in Montreal? _____

_____

_____

_____

Oymyakon—Station # _____

9. Briefly discuss the adaptations that you think are required to live in climates that experience these winter conditions (transportation, housing, outdoor activities). _____

_____

_____

_____

10. Explain why this station experiences this climate—contributing factors.

_____

_____

_____

✳ SECTION 4

The Desert Climates

The *B* climates are the only ones that Köppen classified according to the *amount and annual distribution of precipitation*. POTET (potential evapotranspiration) exceeds PRECIP (precipitation) in all parts of *B* climates, creating varying amounts of permanent water deficits that distinguish the different subdivisions within this climatic group. Vegetation is typically xerophytic, that is, drought resistant, waxy, hard leafed, and adapted to aridity and low transpiration loss. In general terms the major subdivisions are the *BW deserts*, where PRECIP is *less* than one-half of POTET,

and the *BS steppes*, where PRECIP is *more* than one-half of POTET.

In an effort to better approximate the dry climates, Köppen developed simple formulas, graphically presented in Figure 11.3, to determine the usefulness of rainfall, based on the season in which it falls—whether it falls principally in the summer with a dry winter, or in the winter with a dry summer, or whether the rainfall is evenly distributed throughout the year. Winter rains are the most effective because they fall at a time of lower POTET.

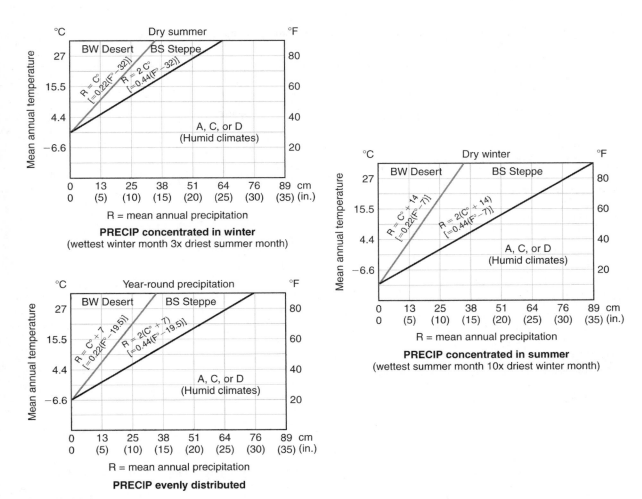

Figure 11.3
B climate determination

As in ✳ SECTION 2, complete two climographs (Figures 11.4a and 11.4b). Using the climate classification key presented in Table 11.1 and the three graphs in Figure 11.3 above, determine the climatic classification for each station and the identity of each station—Reno, Nevada, or Walgett, New South Wales. Complete the analysis and information sheet for each of the stations.

　　　　　　　　　　　　　　　　　　　　　　Lab Exercise 11 • **157**

Figure 11.4a

Station #7: _____

Total annual rainfall: _____

Average annual temperature: _____

Annual temperature range: _____

Distribution of temperature during the year:

_____

_____

Distribution of precipitation during the year:

_____

_____

Principal atmospheric lifting mechanism(s):

_____

_____

Distribution of potential evapotranspiration during

the year: _____

_____

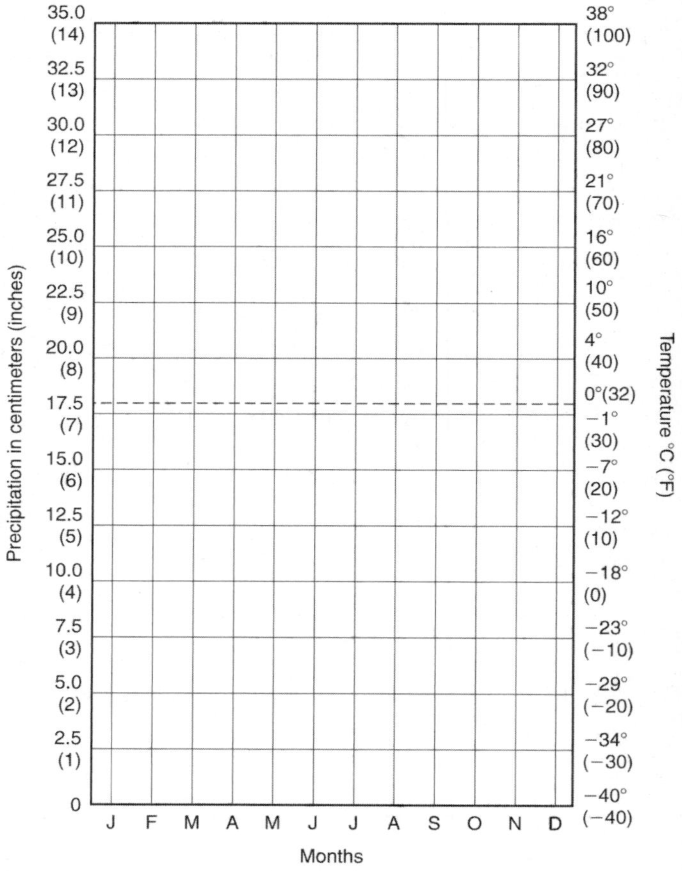

	Jan	Feb	Mar	Apr	May	Jun	Jul	Aug	Sep	Oct	Nov	Dec	Annual
Temperature °C	28.1	27.2	24.4	19.7	15.0	11.7	10.8	12.5	19.2	20.6	24.4	27.2	20.0
(°F)	(82.5)	(81.0)	(76.0)	(67.5)	(59.0)	(53.0)	(51.5)	(54.5)	(66.5)	(69.0)	(76.0)	(81.0)	(68.0)
PRECIP cm	5.3	4.8	4.1	3.0	3.8	4.1	3.3	2.8	2.5	3.0	3.8	4.3	45.0
(in.)	(2.1)	(1.9)	(1.6)	(1.2)	(1.5)	(1.6)	(1.3)	(1.1)	(1.0)	(1.2)	(1.5)	(1.7)	(17.7)
POTET cm	17.8	14.5	11.9	6.6	3.6	1.8	1.8	2.5	6.3	8.4	12.7	17.3	104.6
(in.)	(7.0)	(5.7)	(4.7)	(2.6)	(1.4)	(0.7)	(0.7)	(1.0)	(2.5)	(3.3)	(5.0)	(6.8)	(41.2)

What are the main climatic influences for this station (air pressure, air mass sources, degree of continentality, temperature of ocean currents)? _____

_____

_____

_____

Köppen climate classification symbol: _____; name: _____

_____explanation for this determination: _____

_____

_____

Representative biome (terrestrial ecosystem) characteristic of region: _____;
characteristic vegetation: _____

Figure 11.4b

Station #8: _____

Total annual rainfall: _____

Average annual temperature: _____

Annual temperature range: _____

Distribution of temperature during the year:

_____

_____

Distribution of precipitation during the year:

_____

_____

Principal atmospheric lifting mechanism(s):

_____

_____

Distribution of potential evapotranspiration during

the year: _____

_____

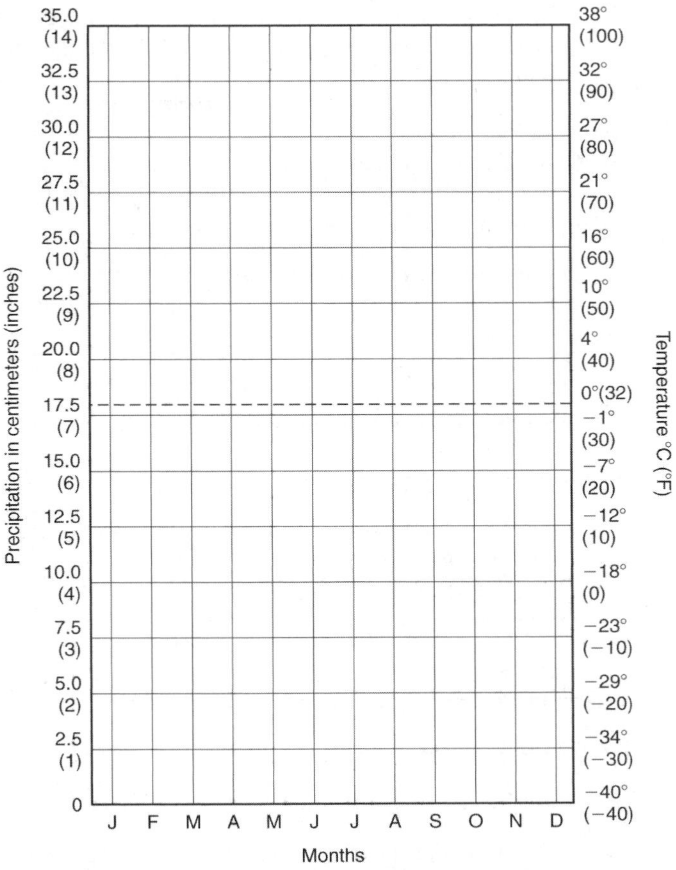

	Jan	Feb	Mar	Apr	May	Jun	Jul	Aug	Sep	Oct	Nov	Dec	Annual
Temperature °C	−0.6	2.2	5.0	8.9	12.8	16.7	21.1	19.4	15.6	10.6	4.4	0.6	10.0
(°F)	(31.0)	(36.0)	(41.0)	(48.0)	(55.0)	(62.0)	(71.7)	(67.0)	(60.0)	(51.0)	(40.0)	(33.0)	(50.0)
PRECIP cm	2.5	2.5	1.8	1.3	1.3	1.0	0.5	0.5	0.5	1.5	1.5	2.3	17.8
(in.)	(1.0)	(1.0)	(0.7)	(0.5)	(0.5)	(0.4)	(0.2)	(0.2)	(0.2)	(0.6)	(0.6)	(0.9)	(7.0)
POTET cm	0	0.7	2.2	4.0	7.1	10.1	13.0	11.7	7.8	4.3	1.7	0.2	62.8
(in.)	(0)	(0.3)	(0.9)	(1.6)	(2.8)	(4.0)	(5.1)	(4.6)	(3.1)	(1.7)	(0.7)	(0.1)	(24.7)

What are the main climatic influences for this station (air pressure, air mass sources, degree of continentality,

temperature of ocean currents)? _____

_____

_____

Köppen climate classification symbol: _____ ; name: _____ ;

explanation for this determination: _____

_____

_____

Representative biome (terrestrial ecosystem) characteristic of region: _____

_____ characteristic vegetation: _____

1. Characterize water balance conditions in **Reno, Nevada** during June, July, and August (compare precipitation and potential evapotranspiration for these months).

_____

_____

_____

2. If you were working in agriculture you might face a shortage of rainfall and need to irrigate plants. Irrigation creates a risk of causing salinization of soils in this semiarid landscape. How would you resolve water

   balance deficits and potential soil problems? Explain. _____

_____

_____

_____

3. What air mass lifting mechanism produces most of the precipitation received in Reno?

_____

_____

_____

_____

4. What air masses produce most of the precipitation received in January and February?

_____

_____

_____

5. What is the annual range of temperature (between January and July) for Reno? _____

   What factors contribute to this value for Reno? _____

_____

_____

_____

6. **Walgett, N.S.W.**, within a temperate short-grass plain region, is in the Southern Hemisphere (**Station #**

   _____). What indications of this global location can you determine from its temperature, PRECIP, and

   POTET data? Explain. _____

_____

_____

7. What primary economic activities are suggested by this climate regime? _____

_____

_____

_____

✳ SECTION 5

Your Climate Region (optional station exercise)

Now that you have examined the various climate patterns throughout the world, you can better understand the climate region within which you are located. Your instructor will give you the necessary climate data for your campus or for a nearby climate station. Use the data to complete the climograph, analysis, and information sheet (Figure 11.5). Note the climate station from SECTION 2 and 4 that is most like your climate. Review the questions about the climate to help you better understand your local conditions.

Station _____

Latitude _____

Longitude _____

Elevation _____

Population _____

Total annual rainfall: _____

Average annual temperature: _____

Annual temperature range: _____

Distribution of temperature during the year:

_____

_____

Distribution of precipitation during the year:

_____

_____

Principal atmospheric lifting mechanism(s):

_____

_____

Distribution of potential evapotranspiration during the year: _____

_____

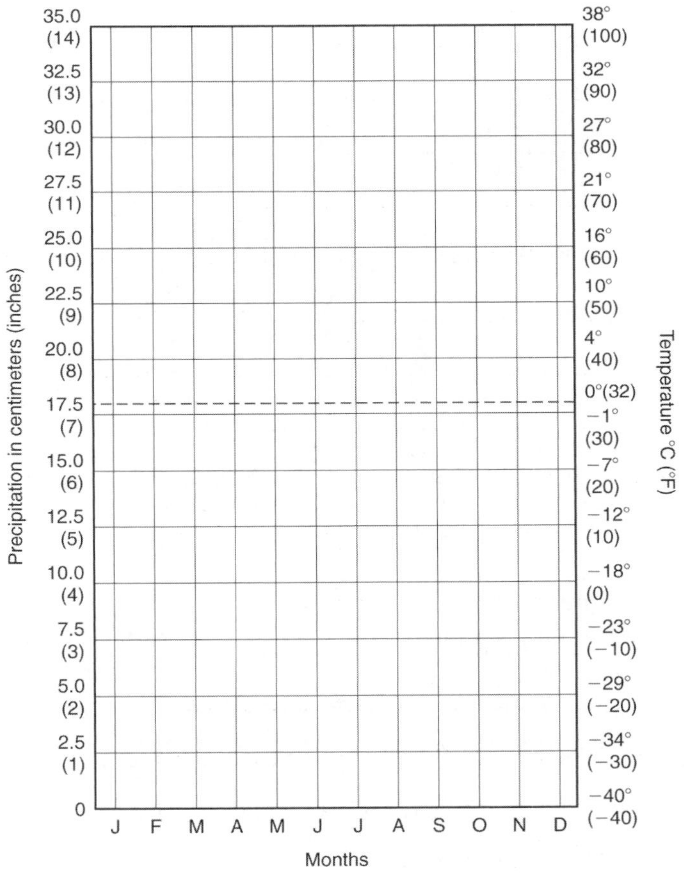

Figure 11.5

Köppen climate classification symbol: _____ ; name: _____ ;

_____ explanation for this determination: _____

_____

_____

Climate Change

Climate regions are not static or fixed, but are dynamic and respond to changes in the environment. As our climate changes, crop patterns, as well as natural habitats of plants and animals, will shift to maintain preferred temperatures. According to climate models for the midlatitudes, climatic regions could shift poleward by 150 to 550 km (90 to 350 mi) during this century. For example, climatic conditions in Illinois are projected to "migrate" southwest as shown in Figure 11.6, that is, climate in Illinois will gradually shift during this century to be more like characteristic climates in east Texas.

Use the climograph (Figure 11.7) and the data that follows to plot current mean monthly temperature in blue (*solid-line graph*), projected mean monthly temperature in red (*solid-line graph*), current mean monthly precipitation in blue (*a bar graph*), projected mean monthly precipitation in red (*a bar graph*), current potential evapotranspiration in blue (*a dotted line*), data and projected potential evapotranspiration in red (*a dotted line*) data for the following station. You may have to color in just the outlines of the precipitation bars so you will be able to see both sets of data on the same climograph. Use Table 11.1 to determine the station's current and projected climatic classification. Additionally, use the map of Earth's Terrestrial Biomes on the inside back cover of this manual to establish the current and projected biomes.

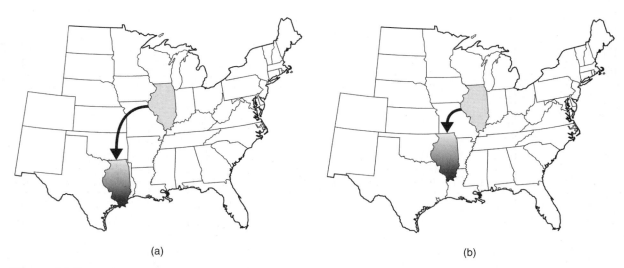

(a)   (b)

Figure 11.6
Migrating summer and winter Illinois climates from 2005 to 2090.
These are climatic projections based on moderate CO_2 emission scenarios.
After "Climate Change Projections for the United States Midwest" by D. Wuebbles and K. Hayhoe, *Mitigation Strategies for Global Change* 9: 335–363, 2004. Kluwer Academic Publishers.

1. **Cairo, IL:** _____

   Latitude 38.05° N
   Longitude 89.18° W
   Elevation 310 ft
   Population 3,632
   Current and projected total annual rainfall:

   _____

   Current and projected average annual
   temperature: _____ C 14.6   P 20.4 _____

   _____

   Current and projected annual temperature range:
   _____ C 26.2 _____ P 30.2 _____

   _____

   Current and projected distribution of temperature during the year:

   _____

   Current and projected distribution of precipitation during the year:

   _____

   Current and projected distribution of potential
   evapotranspiration during the year:

   _____

   _____

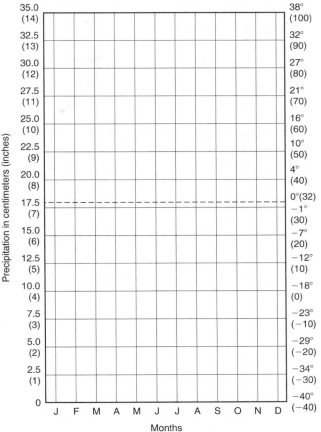

Figure 11.7

Current	Jan	Feb	Mar	Apr	May	Jun	Jul	Aug	Sep	Oct	Nov	Dec	Annual
Temperature °C	0.7	3.6	9.1	15	20.3	24.9	26.9	25.7	21.7	15.5	8.8	3.3	14.7
(°F)	(33.4)	(38.6)	(48.4)	(59.1)	(68.6)	(76.8)	(80.5)	(78.2)	(71)	(59.9)	(47.9)	(37.9)	(58.4)
PRECIP cm	8.2	9	11.2	12	12.1	10.5	11.1	9.2	7.7	8.7	11.2	10.7	121.5
(in.)	(3.2)	(3.5)	(4.4)	(4.7)	(4.8)	(4.2)	(4.4)	(3.6)	(3.0)	(3.4)	(4.4)	(4.2)	(47.8)
POTET cm	0.5	0.8	2.8	6.6	11	14.8	16.8	15.1	10.8	6.4	2.3	0.7	88.7
(in.)	(0.2)	(0.3)	(1.1)	(2.6)	(4.3)	(5.8)	(6.6)	(5.9)	(4.3)	(2.5)	(0.9)	(0.3)	(34.9)

Projected	Jan	Feb	Mar	Apr	May	Jun	Jul	Aug	Sep	Oct	Nov	Dec	Annual
Temperature °C	4.7	7.6	14.1	20.0	25.3	32.9	34.9	33.7	27.7	21.5	14.8	7.3	20.4
(°F)	(40.6)	(45.8)	(57.4)	(68.1)	(77.6)	(91.2)	(94.9)	(92.6)	(81.8)	(70.7)	(58.7)	(45.1)	(68.7)
PRECIP cm	7.4	8.1	12.9	13.8	13.9	8.3	8.8	7.3	8.2	9.3	12.0	9.6	119.6
(in.)	(2.9)	(3.2)	(5.1)	(5.4)	(5.5)	(3.3)	(3.5)	(2.8)	(3.2)	(3.6)	(4.7)	(3.8)	(47.0)
POTET cm	0.9	0.9	4.3	7.5	12.7	16.8	18.4	17.4	13.6	8.8	2.8	1.2	105.3
(in.)	0.4	0.4	1.7	3.0	5.0	6.6	7.2	6.9	5.4	3.5	1.1	0.5	41.5

2. Current Köppen climate classification symbol:_____; name: _____

_____explanation for this determination: _____

_____

_____

Projected Köppen climate classification symbol: _____; name: _____

_____explanation for this determination: _____

_____

_____

3. You can see that the climate categories are broad and cover differing conditions within a single classification. Do you think that the Köppen climate classification accurately describes the amount of change in temperature, precipitation, and POTET between the present and projected conditions? Explain.

_____

_____

_____

4. Compare the position of Cairo, IL in Figure 11.6 to the world climate map on the back cover of this lab manual. What climate classification border is closest to Cairo, IL under current and projected conditions? At present you find the city near the microthermal–mesothermal boundary. Describe its future relations to

conditions on this map. Present location: _____

_____

Projected location: _____

_____

5. Using the world biome map inside the back cover of this lab manual or the one in your textbook determine:

Current representative biome (terrestrial ecosystem) characteristic of the region: _____

_____; characteristic vegetation: _____

Projected representative biome (terrestrial ecosystem) characteristic of the region: _____

_____; characteristic vegetation: _____

Imagine the impact on agriculture and the soils and soil processes on which crops are dependent. The key question for our future is: As temperature patterns change, how fast can plants either adapt to new conditions or migrate to remain within their shifting specific habitats? Rapid changes in vegetation patterns in northern latitudes since 1980 are indicated from satellite observations. Global warming is reducing spring snow cover, allowing spring greening to occur earlier, up to three weeks sooner in some high latitude locations. Likewise the onset of fall is delayed to a later date than previously experienced. Take a moment and speculate on what the impacts will be on the following:

a) Plant respiration increases and net photosynthesis, since a lot of the warming is in nighttime

temperatures: _____

_____

_____

b) River and reservoir levels and water resources (including impacts on water transport): _____

_____

_____

c) Available soil moisture in soil moisture storage: _____

_____

_____

_____

d) Disease vectors, such as mosquitos: _____

_____

_____

6. Under moderate climate change scenarios, summer heat index temperatures in the 21st century are expected to rise in the American southeast in a range between 2.7 C° and 13.8 C° (5 F° and 25 F°). How do you think this will affect air conditioning demand and related costs? After reading either Focus Study 5.1 in *Geosystems*, sixth edition, or Focus Study 3.2 in *Elemental Geosystems*, fourth edition, what are your concerns and what actions would you take if you were a Chicago public health official?

_____

_____

7. Take a moment and brainstorm with lab colleagues your proposals to slow these changes in climate characteristics to allow more time for society to adapt. From *Geosystems* or *Elemental Geosystems,* have you learned of any strategies to slow the greenhouse gas emissions and loading of the atmosphere? Please discuss

and explain: _____

_____

_____

_____

_____

Name: _____   Laboratory Section: _____

Date: _____   Score/Grade: _____

Lab Exercise 12

Plate Tectonics: Global Patterns and Volcanism

The **plate tectonics** theory was a revolution in twentieth-century Earth science. The past few decades have seen profound breakthroughs in our understanding of how the continents and oceans evolved, why earthquakes and volcanoes occur where they do, and the reasons for the present arrangement and movement of landmasses. One task of physical geography is to explain the *spatial* implications of all this new knowledge and its effect on Earth's surface and society.

As Earth solidified, heavier elements slowly gravitated toward the center, and lighter elements slowly welled upward to the surface, concentrating in the crust. Earth's interior is highly structured, with uneven heating generated by the radioactive decay of unstable elements.

The results of this heating and instability are irregular patterns of moving, warping, and breaking of the crust.

In this lab exercise, we work with plate tectonics and the evolution of the present configuration of continents and ocean basins. A reality of plate tectonics is that pieces of Earth's crust are on the move, actively migrating and interacting. Plate interactions are characterized and portrayed in this lab. We examine the correlation among types of plate boundaries and earthquake and volcanic activity. The motion of the Pacific plate and the formation of the Hawaiian Islands are analyzed. Types of volcanic landforms are examined. Lab Exercise 12 features seven sections and includes four optional Google Earth™ activities.

Key Terms and Concepts

asthenosphere
continental drift
core
crust
effusive eruption
hot spots
mantle
mid-ocean ridges
orogenesis

Pangaea
plate tectonics
plumes
sea-floor spreading
seismic waves
subduction zone
transform faults
shield volcano

Objectives

After completion of this lab, you should be able to:

1. *Diagram* the structure of the lithosphere (crust and uppermost mantle).
2. *Analyze* the arrangement of Earth's crustal plates in their former positions.
3. *Contrast* types of plate boundaries and *locate* examples of each type as they are at present.
4. *Analyze* the movement of the Pacific plate and the formation of the Hawaiian-Emperor islands chain.
5. *Analyze* the Kīlauea Caldera, Hawai'i, and Shiprock, New Mexico, volcanic landscapes.

Materials/Sources Needed

color pencils
length of string
scissors

transparent tape
stereolenses

Lab Exercise and Activities

✳ SECTION 1

Earth's Internal Structure and the Crust

As Earth solidified, gravity sorted materials by density. Heavier substances such as iron gravitated slowly to its center, and lighter elements such as silica slowly welled upward to the surface and became concentrated in the crust. Consequently, Earth's interior is sorted into roughly concentric layers, each one distinct in either chemical composition or temperature. Heat energy migrates outward from the center by conduction and by physical convection in the more fluid or plastic layers in the **mantle** and nearer the surface.

Our knowledge of Earth's *internal differentiation* into these layers is acquired entirely through indirect evidence, because we are unable to drill more than a few kilometers into Earth's crust. There are several physical properties of Earth materials that enable us to approximate the nature of the interior.

For example, when an earthquake or underground nuclear test sends shock waves through the planet, the cooler areas, which generally are more rigid, transmit these **seismic waves** at a higher velocity than do the hotter areas, where seismic waves are slowed to lower velocity. This is the science of *seismic tomography*, which allows us to analyze Earth's internal structure just as if Earth was subjected to a kind of CAT scan with every earthquake.

Density also affects seismic-wave velocities. Plastic zones simply do not transmit some seismic waves; they absorb them. Some seismic waves are reflected as densities change, whereas others are refracted, or bent, as they travel through Earth. Thus, the distinctive ways in which seismic waves pass through Earth and the time they take to travel between two surface points help seismologists deduce the structure of Earth's interior (Figure 11.2 in *Geosystems*, eighth edition, or Figure 8.2 in *Elemental Geosystems*, sixth edition).

Completion and analysis items concerning Earth's internal structure.

1. Use the space in Figure 12.1 to sketch Earth's interior from the **core** to the **crust**. Consult a physical geography text, such as *Geosystems*, and lab discussion, to fill in the following labels (name, density, depth) at the appropriate locations. Roughly approximate the scale of each interior layer's thickness in your drawing. (scale: 1 cm = 375 km) After completing your sketch and labeling it, complete the following.

2. Describe the structure of the lithosphere. What layers does it include—from lower to the surface?

_____

_____

3. How have scientists established this portrait of Earth's interior since no one has ever sampled the interior below the crust directly? Explain.

_____

_____

Plates and Plate Boundaries

Figure 12.4 shows tectonic features in modern times (the late Cenozoic Era). A spreading center extends down the center of the Atlantic and into the southern Indian Ocean. Subduction trenches are active along the west coasts of Central and South America and throughout the western Pacific Ocean basin. The northern reaches of the India plate have underthrust the southern mass of Asia through subduction, forming the Himalayas in the upheaval created by the collision.

Of all the major plates, India traveled the farthest. From approximately 150 million years ago until the present, according to Robert Dietz and John Holden, more than half of the ocean floor was renewed. The three inset blocks demonstrate specific types of plate boundaries: *convergent* (subduction), *divergent* (sea-floor spreading, mid-ocean ridges), and *transform* (lateral motion between crests of spreading ridges produces **transform faults**).

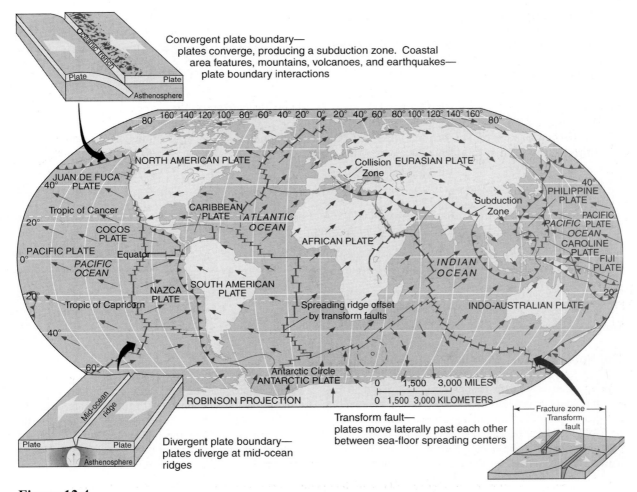

Figure 12.4

The continents today and types of plate boundary interactions.

Each arrow represents 20 million years of plate movement. The longer arrows indicate that the Pacific and Nazca plates are moving more rapidly than the Atlantic plates. See *Geosystems*, eighth edition, Figure 11.18 or *Essential Geosystems*, sixth edition, Figure 8.16, for a map and legend of all the symbols used along plate boundaries.

Answer the following questions and completion items about Earth's plates

1. Using a physical geography textbook, such as *Geosystems*, locate the plate names, examine the different rates of plate movement, and identify types of plate boundary interactions. Using this same source, complete the following questions.

2. Of the three types of plate boundary interactions shown in Figure 12.5, characterize the following plate boundaries:

a) On the ocean floor south of Alaska: _____

b) Indian Ocean between the Antarctic and Indo-Australian plates: _____

c) Beneath the Red Sea: _____

d) Along the west coast of Central America: _____

e) Through Iceland: _____

f) Along the northeast coast of the Persian Gulf: _____

3. Which three plates converge on Japan?

_____

4. Mark your location on the map in Figure 12.4. On which plate do you live? (If you live west of the San Andreas fault system in California, take care with your answer.)

_____

5. Google Earth™ activity, San Andreas Fault, CA. For the kmz file and questions go to mygeoscienceplace.com. Then, click on the cover of *Applied Physical Geography: Geosystems in the Laboratory.*

✳ SECTION 4

Hot Spots and the Hawaiian-Emperor Island Chain

A dramatic aspect of plate tectonics is the estimated 50 to 100 **hot spots** across Earth's surface. These are individual sites of **plumes** of upwelling material from the mantle. Hot spots occur beneath both oceanic and continental crust and appear to be deeply anchored in the mantle, tending to remain fixed relative to migrating plates.

An example of an isolated hot spot is the one that has formed the Hawaiian-Emperor Islands chain. The Pacific plate has moved across this hot, upward-erupting plume for almost 80 million years, with the resulting string of volcanic islands moving northwestward away from the hot spot. Thus, the age of each island or seamount in the chain increases northwestward from the island of Hawai'i.

The big island of Hawai'i actually took less than 1 million years to build to its present stature. Hawai'i Volcanoes National Park is on the island of Hawai'i where volcanic activity is presently nearly continuous. The youngest island-to-be in the chain is still a *seamount*, a submarine mountain that does not reach the surface. It rises 3350 m (11,000 ft) from the ocean floor, but is still 975 m (3200 ft) beneath the ocean surface. Even though this new island will not see the tropical Sun for many years, it is already named Lo'ihi.

The map in Figure 12.5 shows the ages of each portion or island in the chain called the Hawaiian-Emperor Seamount Chain.

Questions about and analysis of **Topographic Map #1** and the Kílauea Caldera.

1. What is the scale of this map? _____

   What is the contour interval? _____

2. What is the north-south length and east-west width of Kílauea Crater?

   in miles: _____

   in kilometers: _____

3. What is the vertical relief (the difference between the highest and lowest elevation) between the volcano observatory at Uwekahuna Bluff on the rim of the crater and the floor of the crater at the 3524-foot bench mark? _____ . Note the tick marks on the contour lines.

4. Within the crater is a "fire pit" named Halemaumau. This pit has varied in size during eruptions. What is the relief from the rim of Halemaumau to the 3412-foot benchmark on the floor of the pit? _____

5. Record the dated lava flows noted on the map in the Kílauea Crater, from oldest to youngest:

   _____

   _____

6. Google Earth™ activity. Kílauea Caldera, HI. For the kmz file and questions go to mygeoscienceplace.com. Then, click on the cover of *Applied Physical Geography: Geosystems in the Laboratory.*

✳ SECTION 7

Shiprock Volcanic Neck

Shiprock is an interesting igneous rock formation in northwestern New Mexico. Magma (molten rock) flowed through a conduit to the surface forming a volcanic cone. When eruptions ceased, the volcanic material eroded away, leaving the jagged lava plume standing starkly above the surrounding landscape—a volcanic neck of basalt. Also, magma flowed along linear cracks from the conduit, radiating outward forming *dikes*. Millions of years of erosion exposed this volcanic neck and radiating ridges composed of fine-grained basalt. To the Navajo Nation, the sacred 518 m (1700 ft) tall formation is the "rock with wings."

Using the stereopair photos in Figure 12.7, examine this volcanic feature. (Refer to the section in the Prologue Lab on the use of these photos and stereolenses.) As you work with this image, complete the following.

1. Based only on the appearance of the surrounding landscape, does Shiprock appear to be composed of

   different rock or similar rock materials? Note the observations that led you to your opinion. _____

   _____

2. Is Shiprock intrusive igneous or extrusive igneous in your opinion? Explain. _____

_____

3. How many radiating dikes do you see? _____

_____

4. Describe the weathering processes, physical and chemical, that produced this volcanic formation.

_____

_____

5. Even though northwestern New Mexico is a desert, do you see any indication of water at work as an agent of

erosion and transportation? Explain your observations from the photo stereopairs. _____

_____

_____

6. Google Earth™ activity, Shiprock, NM. For the KMZ file and questions go to mygeoscienceplace.com.
Then, click on the cover of *Applied Physical Geography: Geosystems in the Laboratory.*

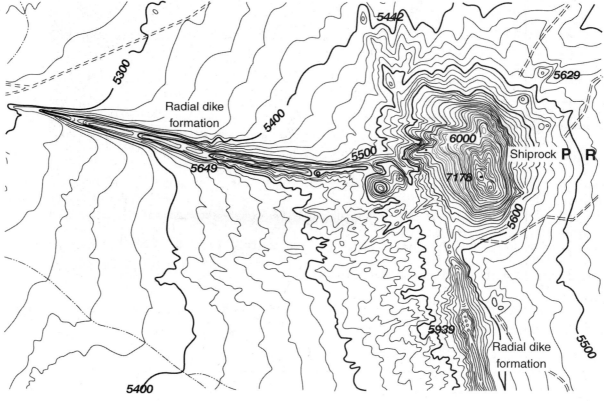

Figure 12.6
Topographic map segment of Shiprock, New Mexico. North is to the right, west to the top. (USGS)

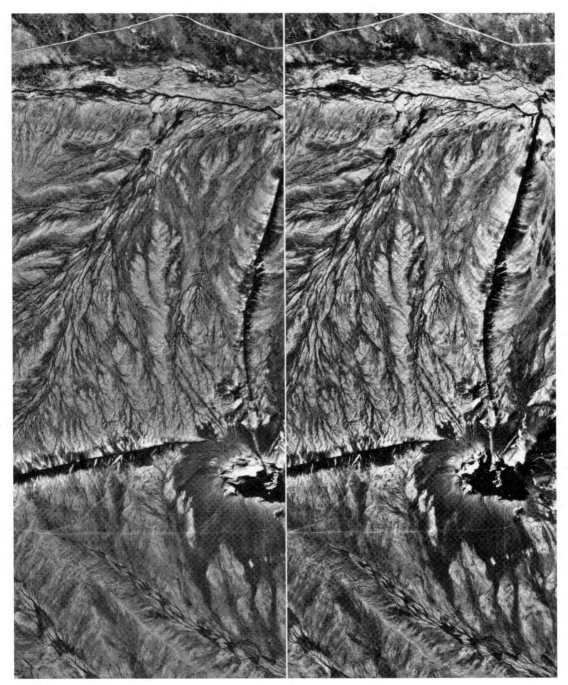

Figure 12.7
Photo stereopair of Shiprock, New Mexico. North is to the right, west to the top. (Photos by NAPP, USGS, Eros Data Center.)

Name: _____   Laboratory Section:_____

Date: _____   Score/Grade: _____

Lab Exercise 13

Recurrence Intervals for Natural Events

Geographers compile spatial data in order to recognize patterns among natural events, make predictions, and encourage preventative or protective actions in an attempt to mitigate the impact of various environmental hazards. Using statistical data to determine probability of the recurrence of a natural event is the focus of this lab exercise. Probabilities are important to guiding protective policies, planning, and hazard zoning, if allowed by the usually sluggish political process. In this lab exercise, we focus on stream processes and the recurring floods that endanger society.

Hydrology is the science of water, its global circulation, distribution, and properties, specifically water at and below Earth's surface. For hydrology links on the web, see http://www.worldwater.org/links.htm, or the Global Hydrology and Climate Center at http://www.ghcc.msfc.nasa.gov/ provides us with a good model for studying recurrence intervals. A recurrence interval, or *return period*, is the average time period within which a given event will be equaled or exceeded once. This exercise examines annual streamflow data to establish recurrence intervals for flooding and to see how this can be used for improved planning and disaster management. Lab Exercise 13 has four sections.

Key Terms and Concepts

discharge
flood frequency curve
flood probability
gage height (flood stage)
hydrology

rating curve
recurrence interval (return period)
stage
streamflow

Objectives

After completion of this lab, you should be able to:

1. *Define* stream discharge and gage height and *describe* the relationship between them.
2. *Use* selected hydrographic data to construct rating curves to *illustrate* streamflow characteristics.
3. *Define* recurrence intervals and use hydrographic data to *calculate* and *graph* them.
4. *Use* rating curves and flood frequency curves to *assess* risk potential and *make* land-use decisions based on flood potential.

Materials/Sources Needed

pencils
calculator
color pencils
ruler
Internet to obtain local hydrologic data (optional in Section 3)

Lab Exercise and Activities

✳ SECTION 1

Streamflow, Discharge, and Gage Height

Stream channels vary in *width* and *depth*. The streams that flow in them vary in *velocity* and in the *sediment load* they carry. All of these factors increase with increasing **discharge**, or the stream's volume of flow per unit time. Discharge is calculated by multiplying the velocity of the stream by its width and depth for a specific cross section of the channel, as stated in the simple expression:

$$Q = wdv$$

where Q = discharge, w = channel width, d = channel depth (measured above an arbitrary datum called the stage), and v = stream velocity. As Q increases, some combination of channel width, depth, and stream velocity increases. Discharge is expressed in cubic meters per second (cms or m^3/sec) or cubic feet per second (cfs or ft^3/sec). Given the interplay of channel width and depth and stream velocity, the cross section of a stream varies over time, especially during heavy floods.

Flood avoidance or management requires extensive measurement of streamflow characteristics and discharge (flow per unit of time).

Table 13.1 contains data from 1980 to 2009 for the Red River of the North, collected at the gaging station at Grand Forks, North Dakota. (This is only a portion of the data table, which actually contains records since 1882, covering 128 years overall.) The maximum peak discharge and corresponding **gage height** in meters (stream depth above **stage**) are shown for the selected years. Stage is also called stream stage or gage height, and is the height of the water surface, in meters, above an established elevation where the stage is zero. This zero elevation is based on a reference elevation datum, either the National Geodetic Vertical Datum of 1929 (NGVD29) or the later North American Vertical Datum of 1988 (NAVD88). Gage height is also referred to as *flood stage*.

Notes for Table 13.1:
Station name: Red River of the North at Grand Forks, Grand Forks County, ND
Station number: 05082500
Latitude: 47.93° N
Longitude: 97.03° W
Basin name: Sandhill-Wilson

Drainage area (square kilometers): 77,959 km^2

Contributing drainage area (square kilometers): 68,117 km^2
Gage datum elevation: 126 m (above NGVD 29)
Base discharge (cms): 237 cms

Gage heights are given in meters above gage datum elevation.
Discharge is listed in the table in cubic meters per second.
Peak flow data were retrieved from the USGS National Water Information System (NWIS)

Using the data provided in Table 13.1, answer the following questions and completion items.

1. **a)** The highest discharge recorded for the listed years: _____

   b) Year recorded: ____[*1997*]____

   c) The lowest discharge recorded for the listed years: _____

   d) Year recorded: _____

2. **a)** Average discharge (1980–2009): _____

   b) Range of discharge (1980-2009): _____

Table 13.1

Peak annual discharge (in cubic meters per second) and gage height (in meters above gage datum elevation) for the Red River of the North at Grand Forks, North Dakota, 1980–2009. [USGS National Water Information Service, NWIS, http://nwis.waterdata.usgs.gov/usa/nwis/peak.]

1	2	3	4	5
Peak Discharge Date	Discharge (cms)	Gage at Peak	Rank	Recurrence Interval
April 6, 1980	616	9.5	20	1.55
July 1, 1981	188	4.5	28	1.11
April 12, 1982	669	11.3	19	1.63
April 6, 1983	400	8.9	25	1.24
April 2, 1984	904	11.3	13	2.38
May 19, 1985	498	7.9	21	1.48
April 2, 1986	893	11.3	14	2.21
March 29, 1987	490	10.1	23	1.35
April 5, 1988	238	6.4	26	1.19
April 13, 1989	1,109	13.2		4.43
April 5, 1990	141	5.4	29	1.07
July 8, 1991	136	5.4	30	1.03
March 12, 1992	224	7.1	27	1.15
August 3, 1993	734	11.1	18	1.72
July 12, 1994	750	10.5	17	1.82
March 31, 1995	974	12.1	11	2.82
April 21, 1996	1,635	14.0	4	7.75
April 18, 1997	3,836	15.9	1	31
May 21, 1998	832	10.7	16	1.94
March 31, 1999	1,400	13.4	6	5.17
June 26, 2000	882	11.3	15	2.07
April 14, 2001	1,618	13.7	5	6.2
July 13, 2002	1,064	11.6	9	3.4
June 28, 2003	476	7.4	24	1.29
April 1, 2004	960	11.7	12	2.58
June 18, 2005	1,072	12.2	8	3.88
April 6, 2006	2,038	14.6	3	10.33
June 22, 2007	988	11.8	10	3.1
June 16, 2008	496	8.1	22	1.41
April 1, 2009	2,148	15.0	2	15.5

3. a) The highest gage at peak discharge: _____

   b) Year recorded: _____

   c) The lowest gage at peak discharge: _____

   d) Year recorded: _____[1977]_____

4. Scan *Discharge* (column 2) and *Gage at Peak* (column 3) data columns. As gage height at peak (stream depth) gets higher, what usually happens to the discharge? _____

_____

5. In general, there is a direct relationship between discharge and width, depth (gage height), and velocity. A stream's **rating curve** relates stream discharge to the gage height. Using the data in Table 13.1 and the logarithmic graphing paper in Figure 13.1, *plot* the gage height against discharge and draw a smooth-curved line through the plotted points to complete the rating curve. (Note: the numbers on the x-axis (discharge) are multiples of 10. Therefore, the 1983 discharge of 400 cms would be plotted along the x-axis at 40, as can be seen on the graph.) The data for years 1980–1999 have been plotted for you.

6. a) We can use the graph in Figure 13.1 to read data from the chart. Use the rating curve to determine the gage height associated with the following discharges. (Hint: use color pencils to mark them on the graph first.)

     100 cms _____[3.5 m]_____

     500 cms _____

     1000 cms _____

     3000 cms _____

   b) Determine the discharge associated with a gage height (flood stage) of

     6 meters _____[200 cms]_____

     9 meters _____

     12 meters _____

     15 meters _____

7. a) As gage height at peak (stream depth) gets higher, what usually happens to the discharge? _____

_____

   b) Compare the data in Table 13.1 for the following years 1983, 1995, 2005, and 2009. What do you observe in the gage height and discharge relationship in these years?

_____

_____

Stream regulation on the Red River involving storage reservoirs and timed releases downstream from dams can affect the streamflow, resulting in this departure from the statistically average pattern.

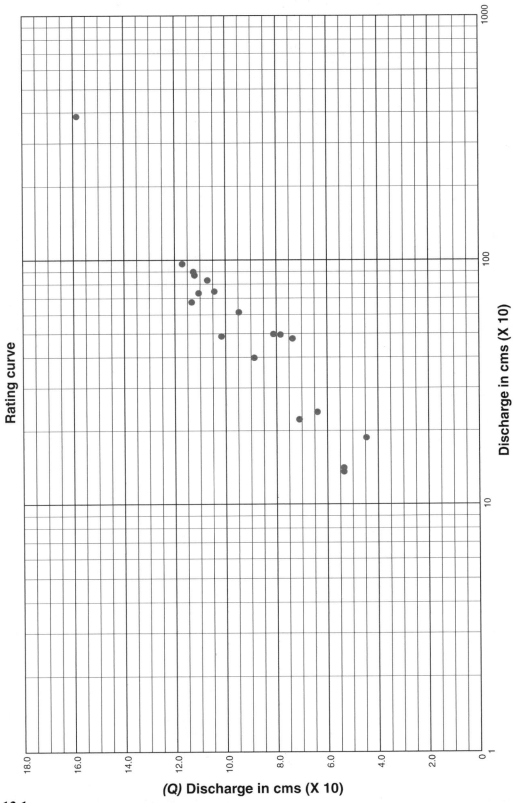

Figure 13.1
Rating curve for the Red River

Floods and Recurrence Intervals

Flood patterns in a drainage basin are as complex as the weather, for floods and weather are equally variable and both include a level of unpredictability. Measuring and analyzing the behavior of each large watershed and stream enables engineers and concerned parties to develop the best possible flood-management strategy. The key to flood avoidance or management is to possess extensive measurements of **streamflow**, a stream's discharge, and how it performs during a precipitation event.

A **flood** is a high water level that overflows the natural river bank along any portion of a stream. Both floods and the floodplains they might occupy are rated statistically for the expected time intervals between floods.

A **recurrence interval**, or *return period*, is the average time period within which a given event will be equaled or exceeded once. A 10-year flood is the highest flood level (**gage height**) that is likely to occur once every 10 years; a 50-year flood is the highest that is likely to occur once every 50 years. A 10-year flood has a 10% likelihood of being equaled or exceeded in any one year and is likely to occur about 10 times in each century. It is important to note that these probabilities estimate *random recurrence*: several decades may pass without experiencing a 10-year flood, whereas floods of such magnitude may occur two or three years in a row. The higher the return interval, the more catastrophic the event is likely to be and the less likely the probability of the event occurring in any given year.

Initial reports on the Midwest floods of 1993 and the Red River floods of 1997 included various references to 300-year and 500-year floods. Data now show that the 1993 Midwest floods easily exceeded a 1000-year **flood probability.** Let us now calculate the recurrence interval for the recent Red River floods.

---

1. Refer once again to Table 13.1 and at the far right locate columns 4 and 5, *Rank* and *Recurrence Interval*, which have not been used as yet. Rank the floods (using the gage height data in column 3), with the rank of 1 for the largest flood and continue in descending order to the smallest, entering your ranking numbers in column 6.

   The recurrence interval (*RI*) is calculated using the following formula:

   $$RI = \frac{n+1}{m}$$

   Where RI = recurrence interval (in years)

   n = number of years of record

   m = rank of the flood

   For example, the flood with a rank of 1 and thirty years of records would have a recurrence interval calculated as follows:

   $$RI = \frac{30+1}{1} = 31$$

2. Calculate the rank and recurrence intervals for the gage heights and enter your results in column 5 of Table 13.1. Some (namely, flood rank 11–30) have been done for you on the table to get started.

3. On the logarithmic paper in Figure 13.2, *plot* the recurrence (return) interval against the discharge and draw a straight line through the plotted points (so that the line fits with approximately half of the dots above and half below the line) to complete the **flood frequency curve.** (Note: the discharge is now on the y-axis and is once again multiplied by 10.) The sample points for flood ranks 1 and 11–30 have been plotted, but the line cannot be drawn until all points are plotted.

Once again you can use the graphs you created to obtain information about hypothetical discharges and recurrence intervals.

4. a) Using the flood frequency curve (and color pencils to mark on graph), estimate the recurrence interval of a flood of the following magnitudes:

200 cms _____

1500 cms _____[4 years]_____

3000 cms _____

b) Estimate the discharge of the

50-year flood _____

100-year flood _____

500-year flood _____

5. a) Use Figure 13.2 along with Figure 13.1 to determine the flood stage (gage height) of these flood frequencies. (Use the flood frequency curve in Figure 13.2 to obtain the discharge, and then use that discharge on the rating curve to determine the associated gage height.)

20-year flood _____

300-year flood _____

400-year flood _____

b) Now use both graphs (Figure 13.1 first, and then 13.2) to determine the recurrence interval of the flood stage (gage height) at

10 meters _____

12 meters _____

15 meters _____

6. How many times in a century would you expect the Red River to crest at a gage height of 13 meters?

_____

What is the probability of a flood of that magnitude in any given year? _____

7. According to the flood frequency curve, what is the return interval for a flood the magnitude of the spring

1997 flood? _____

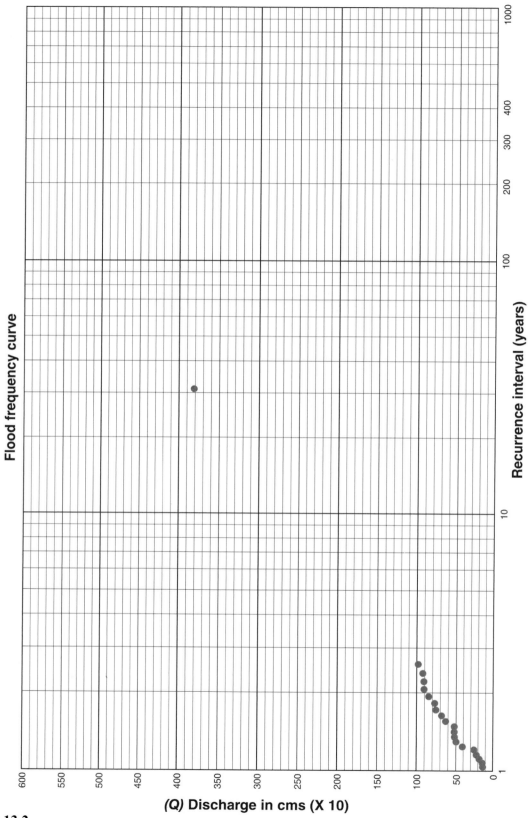

Figure 13.2
Flood frequency curve for the Red River

8. We must be careful in trying to extrapolate risks with a limited set of data. Your frequency curve was based on the last 30 years of data (out of 120 years of record keeping). How do you think using all 120 years of data

would affect your flood frequency curve, if at all? _____

_____

_____

✳ SECTION 3

Comparing Hydrologic Data

In the previous sections of this exercise the Red River of the North served as an example of how hydrologists graph collected data and determine flood frequency. This section contains data tables (Tables 13.2 and 13.3) for gaging stations along two other U.S. streams: the **Missouri River** (at Boonville, Missouri) and the **Willamette River** (at Salem, Oregon). Your instructor may place class members in groups to plot the data for one of these rivers. Complete the graph and questions for your assigned river, then compare with a group that worked with the other river. There is graph paper and completion sections for one river provided here.

Data for streams in the United States are available on the US Geological Survey's website (http://nwis.waterdata.usgs.gov/usa/nwis/peak).

Your instructor may choose to obtain data for streams in your local area to use in this section. (*Note: If you obtain your own local data, remember to convert the English units into metric units: Use the conversion chart on the fold-out flap of this lab manual.*)

Following the example in ✳ SECTION 1, once again answer similar questions using data from one of the following three tables for the river on which you are working (or data from a local stream). Put your answers below. Plot peak discharge and stream gage height on the rating curve graph in Figure 13.3. Compare your results with the Red River data and data from the other streams. What differences do you see? Discuss this with others in the lab.

Complete the following relative to the rating curve for the _____ River.

1. a) The highest discharge recorded for the listed years: _____

   b) Year recorded: _____

   c) The lowest discharge recorded for the listed years: _____

   d) Year recorded: _____

2. a) Average discharge (1980–2009): _____

   b) Range of discharge (1980–2009): _____

3. a) The highest gage at peak discharge: _____

   b) Year recorded: _____

   c) The lowest gage at peak discharge: _____

   d) Year recorded: _____

4. Examine *Discharge* (column 2) and *Gage at Peak* (column 3) data columns. As gage height at peak (stream depth) gets higher, what usually happens to the discharge? _____

_____

5. a) We can use the graph to read data from the chart. Use the rating curve to determine the gage height associated with the following discharges. (Hint: use color pencils to mark them on the graph first.)

100 cms _____

500 cms _____

1000 cms _____

3000 cms _____

b) Determine the discharge associated with a gage height (flood stage) of

6 meters _____

9 meters _____

12 meters _____

15 meters _____

Now use your new data to rank the floods and calculate return frequencies, entering the figures in the appropriate columns on the data table for your river. Plot those figures on the flood frequency curve in Figure 13.4. Use these new graphs to answer questions similar to those in ✳ SECTION 2. By comparing the various graphs your lab class created, you can see how these streams differ in their flood patterns.

Complete the following relative to the flood frequency curve for the _____ River.

6. a) Using the flood frequency curve (and color pencils to mark on graph), estimate the recurrence interval of a flood of the following magnitudes:

200 cms _____

1500 cms _____

3000 cms _____

b) Estimate the discharge of the

50-year flood _____

100-year flood _____

500-year flood _____

7. a) Use Figure 13.2 along with Figure 13.1 to determine the flood stage (gage height) of these flood frequencies. (Use the flood frequency curve in Figure 13.2 to obtain the discharge, and then use that discharge on the rating curve to determine the associated gage height.)

20-year flood _____

300-year flood _____

400-year flood _____

b) Now use both graphs (Figure 13.1 first, and then 13.2) to determine the recurrence interval of the flood stage (gage height) at

10 meters _____

12 meters _____

15 meters _____

If you have time, your instructor may want to discuss these stream rating and flood frequency patterns for these three rivers and the Red River you worked with earlier. Related factors might include climate regions, drainage area, river valley shape, periodic times of apparent drought or heavy precipitation (El Niño/La Niña years), among others.

General notes for data tables:

Data in the following tables are adapted from "Peak Streamflow for the Nation" listed on the USGS website—http://nwis.waterdata.usgs.gov/usa/nwis/peak. (The original data were reported in English units, but are converted into metric units for use in this exercise.)

• Gage heights are given in meters above gage datum elevation.

• Discharge is listed in the table in cubic meters per second.

• Just as in accounting where a "fiscal year" may differ from the calendar year (July 1–June 30 vs. January 1–December 31), the data-gathering period for streams differs from the calendar year. Thus, in some instances the peak discharge date differs from the actual "calendar" year. Peak flows occurring toward the end of a calendar year may be the peak flow for the following calendar year. For example, in Table 13.2, the Missouri River had two peak flows in 1993: the first one in July was for the 1993 data period, and the second one in October was for the 1994 data period.

• Due to the extremely high potential discharge of the **Missouri River**, the "Q" values must be multiples of 100 rather than 10 on both the rating and flood frequency curves. Mark the axis labels to reflect this. (Be sure to keep this in mind when comparing the patterns of the various rivers.)

Notes for Table 13.2:
Station name: Missouri River at Boonville, Cooper County, Missouri
Station number: 06909000
Latitude: 38.98° N
Longitude 123.05° W

Drainage area (square kilometers): 1,299,403 km^2

Gage datum elevation: 172.34 m (NGVD 29)

Notes for Table 13.3:
Station name: Willamette River at Salem, Marion County, Oregon
Station number: 14191000
Latitude: 44.95° N
Longitude: 92.75° W

Drainage area (square kilometers): 18,855 km^2

Gage datum elevation: 32 m (NGVD 29)

Table 13.2

Peak annual discharge and gage height for the Missouri River at Boonville, Missouri, 1980–2009.

1	2	3	4	5
Peak Discharge Date	**Discharge (cms)**	**Gage at Peak**	**Rank**	**Recurrence Interval**
April 1, 1980	5,320	8.2	23	1.35
July 29, 1981	6,664	9.0	18	1.72
June 11, 1982	7,784	11.3	14	2.21
April 7, 1983	8,896	9.0		
June 13, 1984	7,980	7.6		
February 24, 1985	8,176	4.9		
October 13, 1985	7,924	9.1	11	2.82
October 5, 1986	9,352	9.4		
December 21, 1987	3,500	4.9	27	1.15
September 12, 1989	6,244	9.0	20	1.55
May 17, 1990	8,232	9.7		
May 6, 1991	3,724	7.2	26	1.19
July 28, 1992	5,740	5.7	21	1.48
July 29, 1993	21,140	6.7		
October 1, 1993	7,112	10.1	16	1.94
May 19, 1995	10,108	8.4		
May 29, 1996	8,288	8.6		
April 14, 1997	7,868	5.8	12	2.58
April 2, 1998	6,692	9.2	17	1.82
October 7, 1998	804	8.2	30	1.03
June 26, 2000	3,752	6.5	25	1.24
June 8, 2001	8,428	8.4		
May 14, 2002	6,524	8.1	19	1.63
May 12, 2003	3,164	9.2	28	1.11
August 29, 2004	4,536	5.1	24	1.29
June 14, 2005	5,572	9.3	22	1.41
October 5, 2005	3,136	8.3	29	1.07
May 10, 2007	9,716	9.0		
September 16, 2008	7,868	7.5	13	2.38
April 30, 2009	7,700	8.4	15	2.07

Table 13.3
Peak annual discharge and gage height for the Willamette River at Salem, Oregon, 1980–2009.

1	2	3	4	5
Peak Discharge Date	Discharge (cms)	Gage at Peak	Rank	Recurrence Interval
December 27, 1980	344	7.6	30	1.03
December 7, 1981	3,920	8.2		
February 20, 1983	3,808	8.0		
February 15, 1984	2,996	7.1		
November 30, 1984	2,968	7.1	11	2.82
February 24, 1986	2,629	6.6	17	1.82
November 29, 1986	4,256	8.5		
January 16, 1988	2,792	6.8	15	2.07
January 12, 1989	2,968	7.1	12	2.58
January 9, 1990	2,587	6.5	18	1.72
January 16, 1991	2,304	6.1	23	1.35
December 8, 1991	1,949	5.5	26	1.19
March 24, 1993	2,066	5.7	25	1.24
February 25, 1994	2,338	6.2	22	1.41
January 16, 1995	1,702	5.1	28	1.11
February 8, 1996	3,500	7.7		
January 2, 1997	6,832	10.7		
January 15, 1998	4,760	9.0		
December 30, 1998	2,534	6.5	20	1.55
November 27, 1999	4,536	8.8		
December 24, 2000	2,912	7.0	13	2.38
January 27, 2002	1,000	3.9	29	1.07
February 1, 2003	2,439	6.3	21	1.48
December 15, 2003	2,694	6.7	16	1.94
December 12, 2004	2,554	6.5	19	1.63
January 1, 2006	1,840	5.4	27	1.15
December 27, 2006	3,948	8.2		
December 27, 2006	2,884	6.9	14	2.21
January 13, 2008	2,288	6.1	24	1.29
January 3, 2009	3,136	7.3		

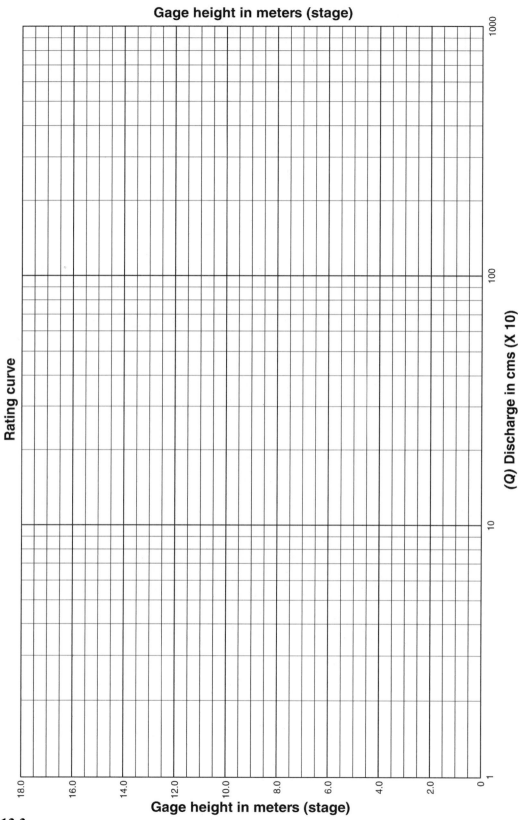

Figure 13.3
Rating Curve for the _____ River

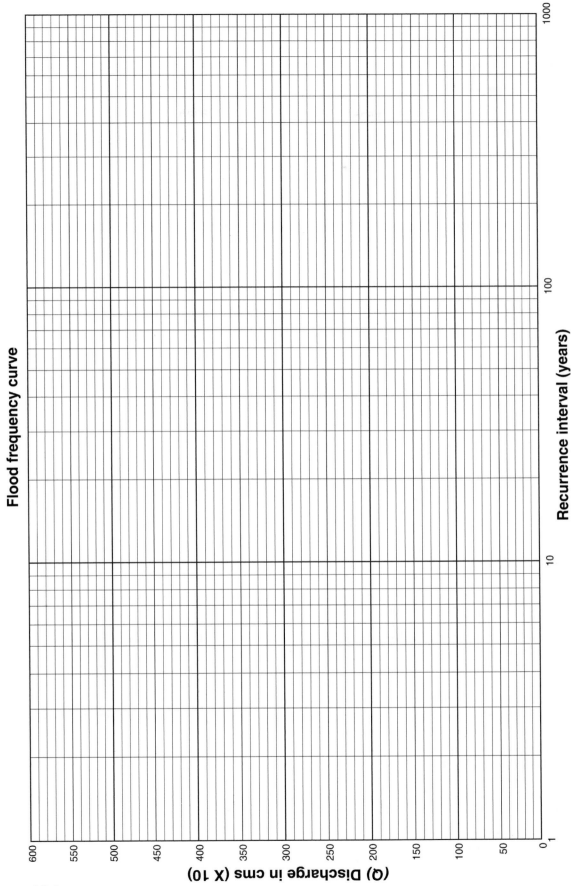

Flood frequency curve

Discharge in cms (X 10) (Q)

Recurrence interval (years)

Figure 13.4
Flood Frequency Curve for the _____ River

Probability, Prediction, and Risk Assessment

Throughout history, civilizations have settled flood-plains and deltas, especially since the agricultural revolution of 10,000 years ago, when the fertility of floodplain soils was discovered. We build artificial levees to keep rivers within their channels and construct dams to hold back and regulate streamflow. How high must the levees and dams be? What type of zoning of the floodplain is required to safeguard settlements? Hydrographic data are compiled and the gage height and flood frequency curves are plotted in order to determine flood risk potential.

Neuse River 1999 Floods

The abuse and misuse of river floodplains brought catastrophe to North Carolina in 1999. In short succession during September and October, Hurricanes Dennis, Floyd, and Irene delivered several feet of precipitation to the state; each storm falling on already saturated ground. With the soil storage at capacity, much of the surplus entered streams and rivers as runoff. About 50,000 people were left homeless and at least 50 died, while more than 4000 homes were lost and an equal amount were badly damaged. The dollar estimate for the ongoing disaster exceeded $10 billion.

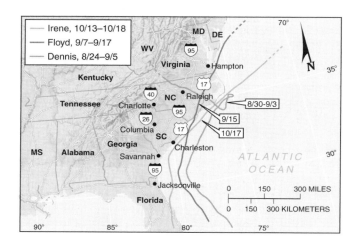

Figure 13.5
North Carolina floodplain disaster. Three hurricanes deluged North Carolina with several feet of rain during September and October 1999, Hurricane Floyd being the worst.

Hogs, in factory farms, outnumber humans in North Carolina. More than 10 million hogs, each producing two tons of waste per year, were located in about 3000 agricultural factories. These generally unregulated operations collect almost 20 million tons of manure into nearly 500 open lagoons, many set on river floodplains. The hurricane downpour flushed out these waste lagoons, spewing hundreds of millions of gallons of untreated sewage into wetlands, streams, and eventually Pamlico Sound and the ocean—a spreading "dead zone." Add to this waste, hundreds of thousands of hog, poultry, and other livestock carcasses, industrial toxins, floodplain junkyard oil, and municipal waste, and you have an environmental catastrophe.

Notes for Table 13.4:
Station name: Neuse River at Kinston, Lenoir County, North Carolina
Station number: 02089500
Latitude: 35.25° N
Longitude: 75.6° W

Drainage area (square kilometers): 6,454 km^2
Gage datum elevation: 3.3 m (above NGVD)

Table 13.4
Peak annual discharge and gage height for the Neuse River at Kinston, NC, 1980–2009.

1	2	3	4	5
Peak Discharge Date	Discharge (cms)	Gage at Peak	Rank	Recurrence Interval
April 1, 1980	276	4.9	17	1.82
August 27, 1981	196	4.2	26	1.19
January 13, 1982	266	4.8	19	1.63
March 28, 1983	426	5.7		5.17
March 31, 1984	305	5.1	13	2.38
February 20, 1985	253	4.7	20	1.55
December 8, 1985	148	3.7	29	1.07
March 9, 1987	521	6.1		10.33
April 24, 1988	104	3.0	30	1.03
May 11, 1989	398	5.6		4.43
April 8, 1990	273	4.8	18	1.72
August 6, 1991	239	4.6	21	1.48
August 20, 1992	300	5.0	14	2.21
January 16, 1993	339	5.2		3.10
March 11, 1994	330	5.2	12	2.58
February 26, 1995	353	5.5		3.44
September 17, 1996	759	7.1		15.50
February 23, 1997	288	5.1	16	1.94
March 17, 1998	468	6.1		7.75
September 22, 1999	1,016	8.4		31.00
October 25, 1999	451	6.0		6.20
April 9, 2001	218	4.5	24	1.29
January 31, 2002	196	4.3	25	1.24
April 17, 2003	336	5.4	11	2.82
August 24, 2004	218	4.6	23	1.35
March 24, 2005	187	4.2	28	1.11
June 24, 2006	297	5.1	15	2.07
November 30, 2006	367	5.5		3.88
April 12, 2008	194	4.3	27	1.15
March 8, 2009	225	4.5	22	1.41

1. Using Table 13.4 and Figure 13.6, complete the rating curve for the Neuse River at Kinston. The sample points for flood ranks 1 and 11–30 have been plotted for you.

2. Calculate the recurrence intervals for the gage heights and enter your results in column 5 of Table 13.4.

3. On the logarithmic paper in Figure 13.7, *plot* the recurrence (return) interval against the discharge and draw a straight line through the plotted points (so that the line fits with approximately half of the dots above and half below the line) to complete the flood frequency curve. (Note: the discharge is now on the y-axis and is once again multiplied by 10.)

4. The Neuse River reaches flood stage at a gage height of 4.26 m, moderate flood stage at a gage height of 5.48 m, and reaches major flood stage at a gage height of 6.4 m. Flood stage indicates that the river has topped its banks, moderate flooding requires some evacuation and road closures, while major flooding requires evacuation of people and livestock, as well as major road closures.

   a) Using the flood frequency curve from Figure 13.7 (and color pencils to mark on graph) estimate the recurrence interval of a flood of the following magnitudes:

   196 cms _____[1.1 years]_____

   350 cms _____

   518 cms _____

   b) Estimate the discharge of the

   50-year flood _____

   100-year flood _____

   500-year flood _____

5. a) The highest flow recorded on the Neuse River was 1092 cms, recorded in July, 1919. What is the recurrence interval for such a flood, given 86 years of records? _____

   b) Based on the rating curve, what would the gage height be for a discharge of 1092 cms? _____

6. What is the recurrence interval for the 1999 flood event? _____

   Given this interval, what recommendations would you make for managing current and future land use in the

   Neuse River floodplain? _____

   _____

   _____

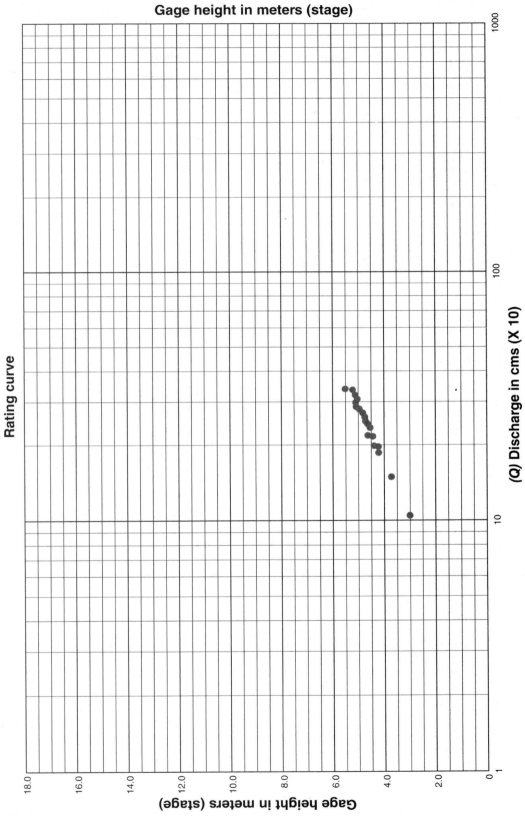

Figure 13.6
Rating Curve for the Neuse River

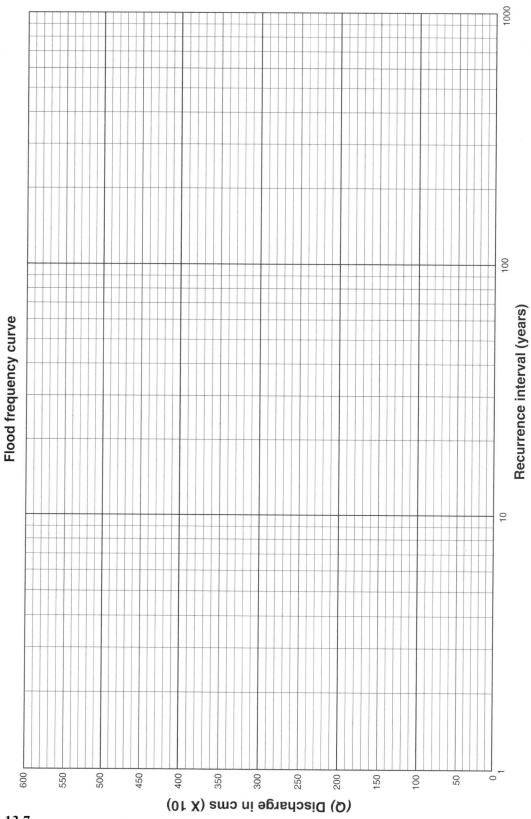

Figure 13.7
Flood Frequency Curve for the Neuse River

Red River, Grand Forks

The levees along the Red River at Grand Forks are currently 15.5 m (50.8 ft) high. In one 1997 event the gage height at peak discharge measured 15.86 m (52 ft), resulting from unusual icing conditions upstream, similar to a "dam break." Four days later the river rose even higher, to its peak gage height for the year, measuring 16.56 m (54.4 ft). Within a few days the river topped the levees twice!

Engineers are raising the Grand Forks levees to 16.76 m (55 ft), including an added lift of 0.91 m (3 ft) to allow for wave action at times of flooding. This is a $300 million project, shared by the citizens of Grand Forks. Ironically, other cities at similar risk along the river have not been as farsighted. This uneven mitigation effort leads to friction and problems between cities and governments when flood events do occur.

Rivers have been important in the geography of human history, influencing where settlements were built, where livelihoods were made, and where borders were drawn. Yet, to what extent should the general public continue to subsidize these risky sites? Rating curves and flood frequency curves can help us make wiser decisions about floodplain settlement and hazard zoning specifications.

7. After reading the above introduction, refer to question 4, ✳ SECTION 2.

   a) What is the recurrence interval for a flood the magnitude of the 1997 flood based on your plotted line? [90 years]

   b) According to your charts, what is the recurrence interval for Grand Forks' peak gage height for that year?

   _____

   c) What are the probabilities of floods of these magnitudes occurring in any given year?

       April 18, 1997 (15.86 m) _____

       April 22, 1997 (16.56 m) _____

   d) What would be the recurrence interval for a flood to reach the top of the new levee? _____

   e) How high above the gage datum would you want to locate a new house in Grand Forks to make certain it

       was safe from a 30-year flood? _____

Lab Exercise 14

Contours and Topographic Maps

An important diagnostic tool for landform analysis is the **topographic map.** While specific application and analysis of topographic maps relating to various types of landforms will be dealt with in later exercises, you have already explored the basic concepts of topographic maps when you analyzed temperature and pressure patterns. **Isolines**, lines connecting points of equal value, are seen on topographic maps as **contour lines,** lines connecting points of equal elevation. These isolines are conceptually similar to the isotherms (equal temperatures) and isobars (equal atmospheric pressure) that you used in Lab Exercises 6 and 7, as well as isohyets (equal precipitation). Therefore, a familiarity with reading and interpreting isolines will be useful at this point. A general discussion of topographic maps and contour lines is the focus of Lab Exercise 14. Lab Exercise 14 features four sections and two optional Google Earth™ activities. These optional questions allow you to compare the topographic maps and your work in the lab manual with Google Earth™ imagery and topographic maps draped over the landscape.

Key Terms and Concepts

Alber's projection
contour interval
contour lines
geographic index number
index contours
isogonic map
local relief
map view
planimetric map

profile view
relief
stereoscopic contour map
topography
topographic map
Universal Transverse Mercator (UTM) grid
vertical datum
vertical exaggeration

Objectives

After completion of this lab, you should be able to:

1. *Construct* contour lines (isolines of equal elevation) and *interpret* a mapped landscape.
2. *Construct* a topographic profile and *calculate* an appropriate vertical exaggeration for a profile.
3. *Describe* and *use* the legend and marginal labels and information on a topographic quadrangle map.

Materials/Sources Needed

pencils
color pencils
calculator
compass
protractor
ruler
stereoscope topographic map of local area

Lab Exercise and Activities

✳ SECTION 1

Contour Lines and Topographic Maps

The USGS depicts information on *quadrangle maps*, rectangular maps bounded by parallels and meridians rather than by political boundaries. A conic map projection, the **Alber's projection** (equal-area), is used as a base for these quadrangle maps (see outside of fold-out flap on back cover of this manual). Two standard parallels (where the cone intersects the globe's surface) are used to improve the accuracy in conformality and scale for the conterminous United States: 29.5° N and 45.5° N latitudes (Figure 14.1). For the U.S. topographic mapping program the standard parallels are shifted for conic projection base maps of Alaska (55° N and 65° N) and Hawai'i (8° N and 18° N).

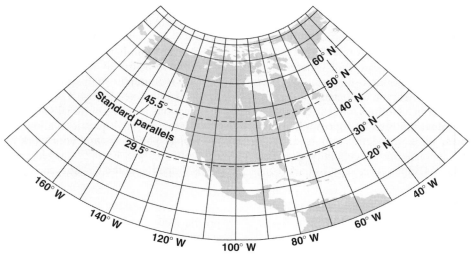

Figure 14.1
Alber's equal-area conic projection with standard parallels of 29.5° N and 45.5° N; a standard for U.S. topographic maps.

In mapping, a basic **planimetric map** is first prepared, showing the horizontal position of boundaries; land-use aspects; and political, economic, and social features. A highway map, such as one that you may have in your car, is a common example of a planimetric map. A vertical scale of physical features is then added to portray the terrain. The most popular and widely used of these maps are the detailed **topographic maps** prepared by the USGS.

Topographic maps portray physical **relief**, the change in elevation between the highest point and lowest point on the map. **Local relief** refers to the change in elevation between two specified points on the map. Relief is indicated through the use of elevation **contour lines**. A contour line connects all points at the same elevation above or below a stated reference level. This reference level—usually mean sea level—is called the **vertical datum**. The **contour interval** is the difference in elevation between two adjacent contour lines. (In Figure 14.2, the contour interval is 20 ft or 6.1 m; in Figure 14.4, the contour interval is 40 ft or 12.2 m.)

More specifics on contour lines:

a) Contour lines are typically printed in *brown color* on topographic maps.

b) All points along a contour line are at the same elevation—a contour line is an *isoline* depicting equal elevation.

c) Contour lines separate areas of higher (upslope) elevation from areas of lower (downslope) elevation.

d) Given a large enough map, contour lines always form a closed polygon although on a smaller map contour lines may run off the margin.

e) The *contour interval* is the difference in elevation between two adjacent contour lines.

f) Contour lines only touch where there is a steep cliff and never cross unless the cliff face has an overhanging ledge (hidden contours are then depicted as dashed lines).

g) Contour lines spaced closer together depict a steeper slope; whereas a wider spacing of contour lines depict a gentler slope.

h) Concentric, closed contours denote a hill or summit; whereas a similar pattern with *hachure marks* (small tick marks) on the downslope side depicts a closed depression.

i) When the map is showing a depression, the first hachured contour line is the same elevation as the adjacent lower contour line.

j) Contour lines form an upstream-pointing, V-shaped pattern wherever contour lines cross a stream.

k) Contour lines occur in pairs from one side of a valley to the other side.

l) Index contours are thicker and usually have their elevation labelled.

The topographic map in Figure 14.2 shows a hypothetical landscape that demonstrates how contour lines and intervals depict slope and relief. Slope is indicated by the pattern of lines and the space between them. The steeper a slope or cliff, the closer together the contour lines appear in Figure 14.2b; note the narrowly spaced contours representing the cliff. A more gradual slope is portrayed by a wider spacing of these contour lines, as you can see from the widely spaced lines on the beach.

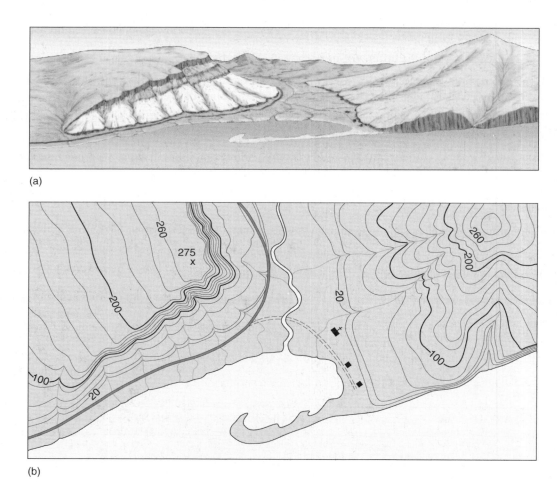

(a)

(b)

Figure 14.2
Perspective view of a hypothetical landscape (a); topographic map of that landscape (b). (After the U.S. Geological Survey.)

Use the hypothetical landscape and topographic map in Figure 14.2 to answer the following questions:

1. What is the contour interval? _____

   How can you tell? _____

2. In terms of local relief (the difference between the highest and the lowest elevation): what is the relief on the

   west (left) side of the highway? _____ on the east (right) side of the highway? _____

3. What is the highest point on the map and what is its elevation? _____

4. Imagine that you are planning a *low-exertion* walk from the church, across the river to the high point on the west side of this landscape. First on Figure 14.2a and then on Figure 14.2b, use a red color pencil to draw the route you would take; give your reasons for choosing this easier route in terms of elevation change per distance traveled.

   _____

   _____

   _____

5. On each figure use a blue pencil to draw in streams, using arrows to show the direction of stream flow. Note there is a main stream in the valley and at least four tributary streams or creeks. How do the contour lines indicate the direction of flow?

   _____

   _____

   _____

Contour lines and contour maps are two-dimensional images portraying relief, which is a three-dimensional concept. A **stereoscopic contour map** (also called a *stereogram*) helps to see these contour lines in three dimensions. Figure 14.3 shows two views of a hill and valley area: (a) is a contour map, and (b) is a stereoscopic contour map.

   Use stereolenses to view Figure 14.3b in three dimensions and compare this view with features marked on the contour map Figure 14.3a.

   A. Each contour line connects points of equal elevation.

   B. Varying degrees of steepness are obvious: the steepest areas where the lines are closest together.

   C. Widely spaced contours depict where the slope is gentle.

   D. Contour lines crossing stream valleys form a V pointing upstream.

   E. Hills are indicated by a series of closed, concentric contours.

   F. Depression contours have hachure marks on the downslope side. The hachures are not marked on the stereo pairs in Figure 14.3b as are on the contour map in Figure 14.3a, but the depression is clearly evident as you view the stereo pair in 3-D through the stereoscope.

Once you become adept at reading contour lines, you will be able to "see" the relief, even though the maps are only in two dimensions. Stereograms will be used again in later lab activities when we examine landforms.

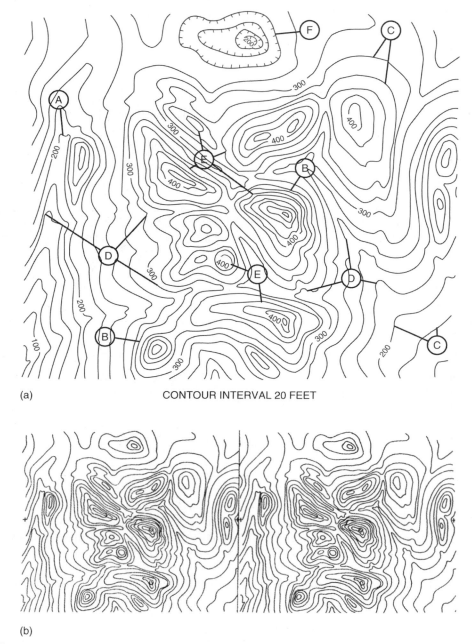

(a)

CONTOUR INTERVAL 20 FEET

(b)

Figure 14.3
Contour map (a) and stereoscopic contour map (b) of the same hill and valley landscape. (From Horace MacMahan, Jr., *Stereogram Book of Contours*. Copyright © 1995, by Hubbard Scientific Company, pp. 8-9. Reprinted by permission of American Educational Products, Hubbard Scientific.)

Constructing Contour Lines

Figure 14.4 presents elevation values for a selected landscape. The contour interval is 40 feet, and **index contours** (thicker lines—on this map every fifth contour, marked with the line's elevation) at 200 feet, 400 feet, and 600 feet are drawn for you in bold isolines. A river channel is noted on the map flowing along a valley separated from the ocean by a ridge. The elevation of the channel is marked at three locations.

Use Figure 14.4 to do the following work and answer the questions.

1. Using a pencil, sketch 40-foot-interval contour lines using the specific site elevations given. Note that we have drawn the 40-ft contour line for you. You must *interpolate* (estimate) elevations between known values. Draw your contour lines lightly at first as you determine the best portrayal, then darken in your work and erase stray pencil marks. (Do not worry if your lines go through the elevation labels.) Remember the contour line basics given in ✳ SECTION 1.

   The coastline is the datum—mean sea level. *Begin your work at the coast and work inland.* (Hint: make this task easier by color-coding the elevations in 40-foot intervals. Circle all elevations from 0-40 feet in green, 41-80 feet in yellow, etc. Then, when drawing the contour lines, you will be "grouping" the colors, with the contour lines as separators between the color groups.)

2. Three vertical control bench marks, indicated by the letters "BM" and marked with an "x," are noted on the map. What are their elevations?

   _____

3. Three other specific spot elevations are noted with an "x." What are their elevations?

   _____

4. What effect do the intermittent (periodically dry) stream channels have on the topography of the region?

   _____

5. Given the trend and location of the 40-foot elevation contour, if you were walking along the water's edge, how would you characterize the topography and relief of the coastline? Steep? Gentle relief?

   _____

6. In what compass direction does the river flow? How can you tell? (See ✳ SECTION 1, question 5.)

   _____

7. Which portion (NW, NE, SE, or SW) of the map has the greatest relief? _____

8. **Challenge question:** If you were assigned the task of building a single general aviation runway of 3700 feet in length, where would you place it? Draw your runway plan on the map to the correct scale. (Make it 100 feet wide.) Give your reasons for this site selection.

   _____

   _____

9. Google Earth™ activity, Stewart's Point Quadrangle. For the KMZ file and questions go to mygeoscienceplace.com. Then, click on the cover of *Applied Physical Geography: Geosystems in the Laboratory*.

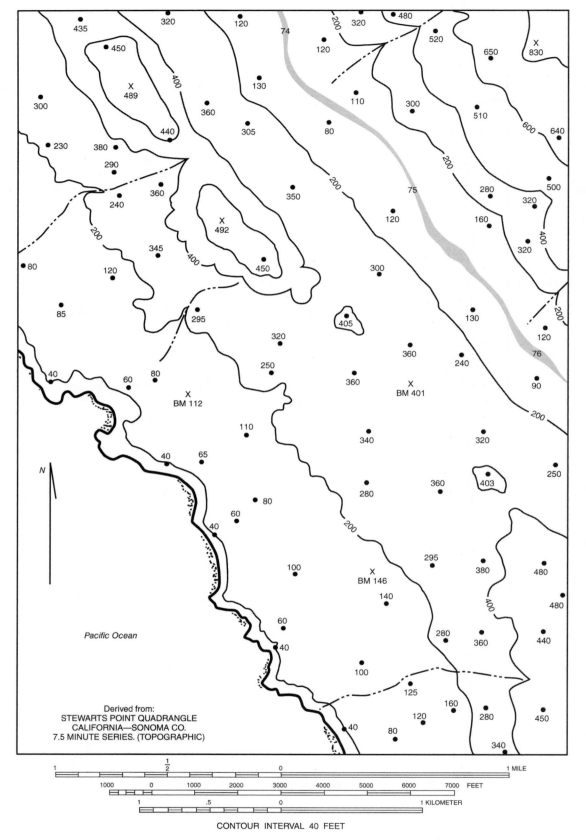

Figure 14.4
A portion of the Stewarts Point quadrangle, enlarged and adapted with 200-, 400-, and 600-foot index contours highlighted.

Topographic Profile

Relief refers to vertical elevation differences in a local landscape. The character and general configuration of Earth's surface is called **topography**, the feature portrayed so effectively on topographic maps. In Figure 14.4 of this exercise you constructed contour lines and completed an analysis of a coastal landscape. The local topography included a coastal marine platform (terrace), low ridge, stream valley, and hill. The maximum relief on the map was approximately 700 feet, averaging about 400 feet.

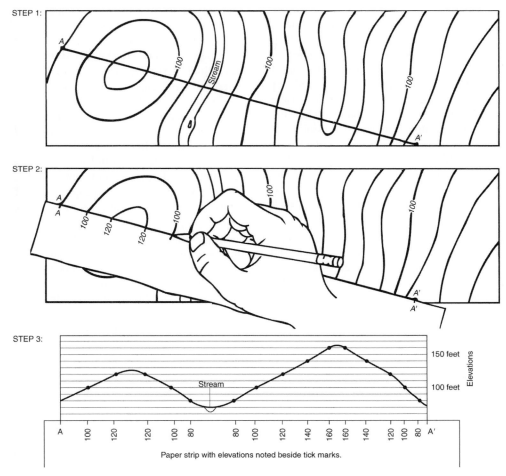

Figure 14.5
Constructing a topographic profile using the edge of a piece of paper. (From Busch, Richard M., editor, *Laboratory Manual in Physical Geology*, 3rd ed., Macmillan Publishing Company © 1993.)

An important method of topographic map analysis is construction of a **topographic profile**, the graphic representation of graduated elevations along a line segment drawn on a map. A **map view**, or plan view, is the normal way we view a map as if we were looking straight down from above ("bird's-eye" view). A cross section of a landscape—as if you sliced through the strata and obtained a side view—is a **profile view** and is demonstrated in Figure 14.5. The profile view shows you the "shape" of the land and demonstrates a line-of-sight perspective. As you work with topographic profiles, allow yourself to develop methods with which you are comfortable.

One method of preparing a profile is to draw a line connecting two points (a *transect*) along which you want to obtain the profile (Figure 14.5, Step 1). Take a piece of paper, fold the paper over and crease it for a stiff straight edge. The paper can be plain or graph paper. (Graph paper divided 10 squares to the inch is included at the back of this manual.) Lay the folded edge along the drawn line. Make tick marks at each point where a contour line makes contact with the edge of the paper and note the elevation at that point. These will represent elevation points on the profile graph (Step 2). Carry the tick marks down to the graph at the appropriate elevation, plotting each

point; connect the points with a smooth curved line to complete the profile (Step 3).

Construct your first topographic profile using Figure 14.6. Portrayed is an enlarged portion of the Palmyra, New York, quadrangle. The western New York region was blanketed during several advances of continental glaciers. The geomorphic feature you are profiling was deposited by a glacier.

A prepared graph is supplied with the Figure; however, let's briefly discuss how it is set up. Note in Figure 14.6 that the horizontal scale of the map is 1:12,000, or $1'' = 888'$. If we used this same scale for the vertical axis on the topographic profile graph, the local relief of 192 feet would only cover 0.192 inch on the graph. To construct a readable and useful profile a technique of **vertical exaggeration** is employed. In this figure a vertical scale of 1:1200, or $1' = 100'$ is used and represents a 10 × (times) exaggeration of the horizontal scale. Other landscapes require different vertical exaggerations—*the greater the maximum relief of a landscape, the greater the scale should be* to conveniently fit on a graph for analysis. For instance, the greater relief in Figure 14.7 made an exaggeration of 3.3×(times) more appropriate as compared to the 10 × exaggeration in Figure 14.6. The horizontal scale should be left at the same scale as the map.

Follow the procedure illustrated in Figure 14.5 to construct a profile along the line drawn in Figure 14.6, making tick marks at each point where the drawn line and contour lines intersect. Carry the elevations denoted by the tick marks down to the graph, plotting each point. The steeper the slope, the closer the points are placed; a gentle slope places them farther apart. The last step is to connect the points with a smooth curved line to visualize the relief and topography of this landscape. You may want to shade the area below the line to better display the profile.

Use the profile constructed in Figure 14.6 to answer the following questions and completion items.

1. What distance does this topographic profile cover? _____

2. What is the maximum relief along this profile? _____

3. The landform feature you have profiled is a drumlin. Drumlins are formed by glacial deposits and are streamlined in the direction of the glacier's movement. They have a blunt end upstream, a tapered end downstream, and a rounded summit. Using your protractor, what direction was the glacier flowing?

_____

4. Using the 500-foot contour, what is the width of this feature? _____

5. If we used a vertical scale identical to the horizontal scale, how many squares on the graph would

accommodate the maximum relief along the profile? _____

6. Google Earth™ activity, Palmyra quadrangle drumlin. For the KMZ file and questions go to mygeoscienceplace.com. Then, click on the cover of *Applied Physical Geography: Geosystems in the Laboratory.*

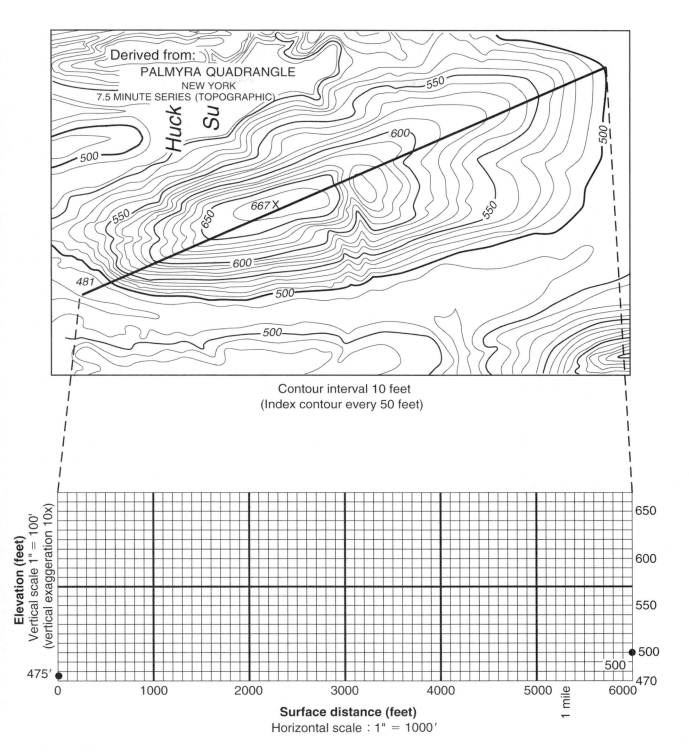

Contour interval 10 feet
(Index contour every 50 feet)

Figure 14.6
Topographic profile from a portion of the Palmyra, New York quadrangle (enlarged).

A second example of a vertical profile is in Figure 14.7. Using the same procedure as you did on the previous assignment, construct a topographic profile. Label the South Fork of the Gualala River (see Figure 14.5, Step 3), the bench mark at 112 feet, and various road crossings on your profile. The coastline forms a contour line of equal elevation—sea level. Note the change in contour interval and scale from that of Figure 14.6.

Use Figure 14.7 to answer further questions about preparing a topographic profile.

7. What is the length of this topographic profile in miles? in kilometers? _____

8. How wide is the marine terrace (a relatively flat platform or "shelf" along the coast) in miles?

_____ in kilometers? _____

9. If we used a vertical scale identical to the horizontal scale, how many squares on the graph would

accommodate the maximum relief along the profile? _____

10. What is the relief (elevation difference between the highest and lowest points, in a local landscape) in the

river valley along southwest slopes? _____

along the northeast slopes? _____

11. Can you physically see the shoreline when standing at the summit at 731 feet elevation? Explain.

_____

_____

Can you see the marine terrace from the summit? _____

Can you see the northwest end of the runway at Sea Ranch Airport (a point just south of the profile) from the

summit? _____ Explain. _____

_____

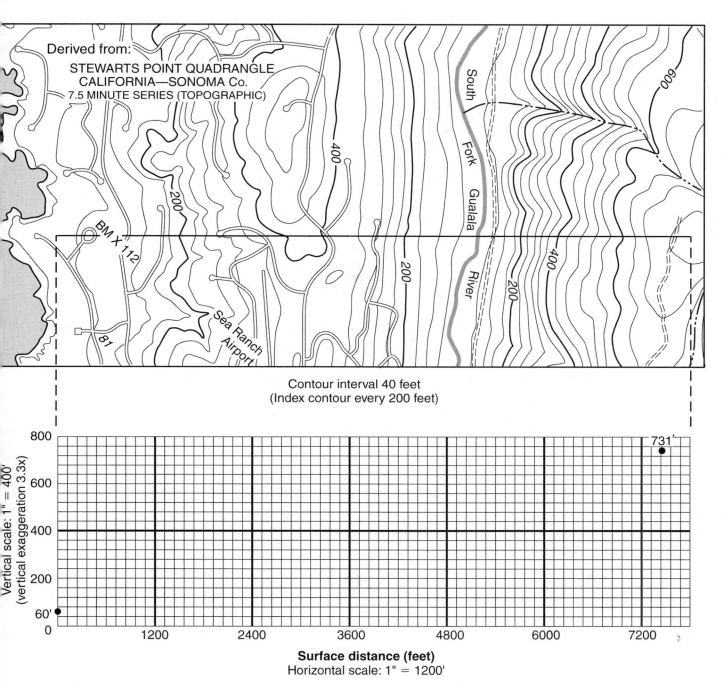

Contour interval 40 feet
(Index contour every 200 feet)

Figure 14.7
Topographic profile from a portion of the Stewarts Point, California quadrangle map (enlarged). From west to east along the profile, this landscape features a coastal marine terrace, a ridge, a river valley, and a hill rising to 731 feet above sea level. The coastline forms a contour line of equal elevation.

Interpreting Topographic Map Information

Topographic maps in the series of quadrangle maps prepared by the USGS, or Centre for Topographic Information in Canada, provide a wealth of information. Have your lab instructor show you an index map for your state or province that portrays all the quadrangle map coverage available. Your lab instructor will provide you with a topographic map sheet to examine as you follow along with this description of its elements.

The name of the topographic map quadrangle is given in the upper-right corner, with the state name, and 15-minute or 7.5-minute series designation; if it is a 7.5-minute quad. This also signifies which portion of the published 15-minute quad it represents ("SW/4, NE/4," etc.). Other grids, in addition to latitude/longitude, are in use, so geographic coordinates of the **Universal Transverse Mercator (UTM)** grid system are listed near the corner margins of the quad map. The Public Lands Survey township and range system used in maps of the western United States is also noted.

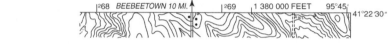

COUNCIL BLUFFS NORTH QUADRANGLE
IOWA–NEBRASKA
7.5 MINUTE SERIES (TOPOGRAPHIC)

The lower-right corner carries the name of the quad; portion of 15-minute quad, if applicable; **geographic index number**; and the year of the map. The *geographic index number* is derived from the latitude and longitude of the lower-right corner of the map. Each corner also features complete geographic coordinates.

COUNCIL BLUFFS NORTH, IOWA—NEBR.

N4115–W9545/7.5

1956

PHOTOREVISED 1969 AND 1975

AMS 6866 IV SE—SERIES V876

The upper-left corner always carries the same credit line for the primary government agent for U.S. mapping.

UNITED STATES
DEPARTMENT OF THE INTERIOR
GEOLOGICAL SURVEY

A more complete credit line and map preparation history is in the lower-left margin. If cooperative assistance from another entity was only minor, this may be listed in this place. A variety of such information might be featured in this label.

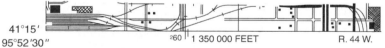

Mapped, edited, and published by the Geological Survey

Control by USGS, USC&GS, and USCE

Topography from aerial photographs by Kelsh plotter
Aerial photographs taken 1952. Field check 1956

Polyconic projection. 1927 North American datum
10,000-foot grid based on Iowa coordinate system, south zone
1000-metre Universal Transverse Mercator grid ticks,
zone 15, shown in blue

Red tint indicates area in which only landmark buildings are shown

Revisions shown in purple compiled from aerial photographs
taken 1969 and 1975. This information not field checked

Purple tint indicates extension of urban areas

Where applicable, the names of adjoining topographic quadrangles are given in parentheses at each corner and along the sides. If this reference appears, you can assume that the map name refers to a quad at the same scale and the same series.

Earth's geographic North Pole and magnetic North Pole locations do not coincide. In addition, the magnetic pole slowly migrates from year to year. The lower left-center margin features a magnetic declination diagram for the year of the map field survey or date of map revision. This declination is taken from an **isogonic map** (Figure 2.4) that maps magnetic declination. This diagram allows you to correct compass bearing readings when using the topographic map. A state map graphically shows the location of the quad in the lower right-center margin. Here are some samples of this designation.

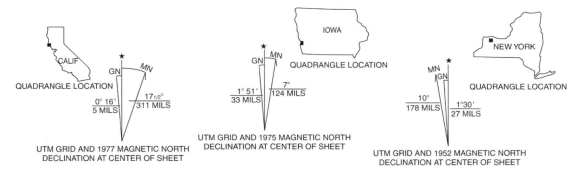

The map scale (representative fraction and graphic or bar scale), contour intervals, any supplementary contour intervals, vertical datum ("National Geodetic Vertical Datum of 1929" appears on post-1975 maps), depth or sounding information in bodies of water, and sales information are shown in the lower-center margin of the map sheet.

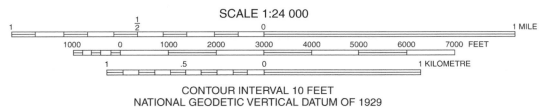

A complete map symbol key should be used for interpreting the map (inside front cover of this lab manual), although road symbols used for the quad are shown in the lower-right corner.

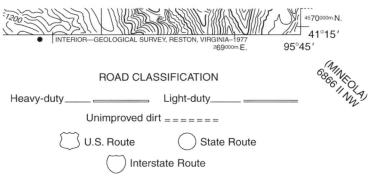

Use a topographic map *provided by your lab instructor* to answer the following questions and completion items.

1. Name, state, and series of the map: _____

_____

2. If a 7.5-minute map, give its location and position relative to the 15-minute series quad for this region:

_____

3. Did any other agencies or entities participate in the preparation of this map?

_____

4. List the geographic index number for this quad:

_____

5. Denote the latitude and longitude of each corner of the quad map:

        N. E. corner: _____

        S. E. corner: _____

        S. W. corner: _____

        N. W. corner: _____

6. List the nearest UTM (Universal Transverse Mercator) coordinate for the upper right corner of the map:

_____

7. Describe the complete map history and credit line:

_____

_____

_____

8. If you needed to obtain the maps adjoining this quadrangle, what are their names? Label these names along the appropriate sides and corners on the quadrangle.

9. What is the representative fraction scale for this map? _____

Convert this to a written (verbal) scale. (Show your work)

_____

10. Using the graphic or bar scale on the map, determine the distance in kilometers between two points selected

by your instructor. The points are: _____ and _____

Distance in kilometers _____; and, miles _____

11. Calculate the total area of the map in square kilometers. (Show your work.) _____

And, in square miles. (Show your work.) _____

12. What is the magnetic declination for your quadrangle map? _____

Which direction (and how many degrees) would you have to turn in order to adjust for this declination?

_____

To what azimuth would your compass needle point? _____

Hikers and backpackers frequently include topographic maps and compasses among their equipment. Magnetic declination must be taken into account when using a compass, and adjustments must be made. To do this, simply superimpose the magnetic compass points (based on the declination arrow) over the true north compass points (N,E,W,S). It is easy to see that, if the magnetic declination is 10°E, you must turn 10° west, so that your compass north arrow is pointing to 350° ($360° - 10° = 350°$).

13. List any additional information provided in the margins of this particular map quadrangle:

_____

_____

_____

Your instructor may want to provide you with additional questions pertaining to the topographic map that has been selected for this activity. Questions may include features specific to the chosen map and use of symbols for which you might need to consult the topographic map legend on the inside front cover of this manual.

Name: _____   Laboratory Section:_____

Date: _____   Score/Grade: _____

Lab Exercise 15

Topographic Analysis: Fluvial Geomorphology

One of the benefits you receive from studying physical geography is an enhanced appreciation of the scenery. As we visit Earth's varied landscapes, we witness the active physical processes that sculpt the land. *Endogenic processes* produce internal forces within our planet that build landforms. However, as the landscape is formed, a variety of *exogenic processes* produce external forces that simultaneously wear these features down. The landscape we see is the product of this continuous struggle of building and destruction.

Geomorphology is a science that analyzes and describes landforms—their origin, evolution, form, and spatial distribution. Geomorphology is an important discipline within physical geography.

Denudation is a general term referring to all processes that cause reduction or rearrangement of landforms. The principal denudation processes affecting surface materials include *weathering, mass movement, erosion, transportation*, and *deposition*, as produced by the agents of moving water, air, waves, and ice, and the pull of gravity.

Fluvial (moving water; overland flow and streams) erosion processes and their resulting landforms dominate land surfaces throughout the world. Specific geomorphic landforms and processes resulting from fluvial actions are analyzed in this exercise using topographic maps and illustrations. Lab Exercise 15 features three sections and three Google Earth™ activities.

Key Terms and Concepts

alluvium
anticline
backswamp
denudation
drainage basins
drainage patterns
floodplain

fluvial
geomorphology
natural levees
oxbow lakes
syncline
watershed
yazoo tributary

Objectives

After completion of this lab, you should be able to:

1. *Analyze* and *describe* fluvial geomorphic processes using topographic maps, photo stereo pairs, and Google Earth™ imagery.
2. *Interpret* several fluvial processes and the characteristic landscapes that are produced, including drainage basins and floodplains.
3. *Analyze* fluvial processes in a folded landscape, Cumberland Gap, Maryland.

Materials/Sources Needed

pencil
ruler
calculator
stereolenses

Lab Exercise and Activities

✳ SECTION 1

Drainage Patterns and Floodplains

Begin this section by reviewing relevant sections in a physical geography text, or *Geosystems*, that covers stream processes and landscapes. Here are a few essentials. Stream-related processes are termed **fluvial** (from the Latin *fluvius*, meaning "river"). Fluvial systems exhibit characteristic processes and produce predictable landforms. The erosive action of flowing water and the deposition of stream-transported materials produce landforms. A stream system behaves with randomness, unpredictability, and disorder as described by the phenomena of chaos. Yet, enough regularity exists to allow some predictability.

A stream is a mixture of water and solids—carried in solution, suspension, and by mechanical transport. **Alluvium** is the general term for the clay, silt, and sand transported and then deposited by running water.

Streams may drain large regions. Consider the travels of rainfall in north-central Pennsylvania. This water feeds hundreds of small streams that flow into the Allegheny River. At the same time, rainfall in southern Pennsylvania feeds hundreds of streams that flow into the Monongahela River. The two rivers then join at Pittsburgh to form the Ohio River. The Ohio flows southwestward and at Cairo, Illinois, connects with the Mississippi River, which eventually flows on past New Orleans into the Gulf of Mexico. Each contributing tributary, large or small, adds its discharge and sediment load to the larger river.

Streams are organized into areas or regions called **drainage basins.** A drainage basin is the spatial geomorphic unit occupied by a river system, defined by ridges that form *drainage divides*, i.e., the ridges are the dividing lines that control into which basin precipitation drains (Figure 15.1).

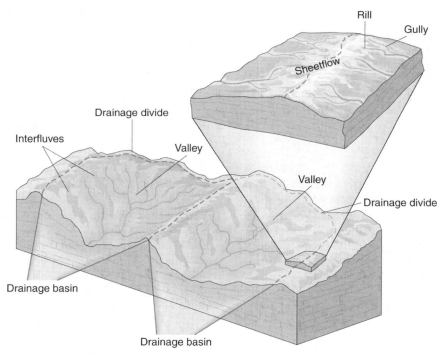

Figure 15.1
The drainage basin and its watershed separated from other basins by a drainage divide.

Drainage divides define a **watershed**, the catchment area of the drainage basin. A major drainage basin system, such as the one created by the Mississippi-Missouri-Ohio river system, is made up of many smaller drainage basins, which in turn comprise even smaller basins; each divided by specific watersheds. Surface runoff concentrates in *rills*, or small-scale indentations, which may develop further into *gullies* and *stream courses* with distinct drainage patterns. The resultant **drainage pattern** is an arrangement of channels determined by slope, differing rock resistance, climatic and hydrologic variability, and structural controls imposed by the landscape.

Figure 15.2 shows seven common drainage patterns. The *trellis* drainage pattern (b) is characteristic of dipping or folded topography, which exists in the nearly parallel Ridge and Valley Province of the eastern United States, where drainage patterns are influenced by rock structures of variable resistance and folded strata. We see this later in this lab in the Cumberland topographic map and photo stereopair in ✱ SECTION 3.

The sketch in Figure 15.2b suggests that the headward-eroding part of one stream could break through a drainage divide and *capture* the headwaters of another stream in the next valley, and indeed this does happen. The sharp bends in two of the streams in the illustration are called *elbows of capture* and are evidence that the stream has breached a drainage divide. This type of capture, or *stream piracy*, occurs in other drainage patterns.

The remaining drainage patterns in Figure 15.2 are caused by other specific structural conditions. *Parallel* drainage (d) is associated with steep slopes. A *rectangular* pattern is formed by a faulted and jointed landscape, directing stream courses in patterns of right-angle turns (e). A *radial* drainage pattern (c) results from streams flowing off a central peak or dome, such as occurs on a volcanic mountain. We will see this later in the Mt. Rainier topographic map in Lab Exercise 16, ✱ SECTION 3. Structural domes, with concentric patterns of rock strata guiding stream courses produce *annular* patterns (f). In areas having disrupted surface patterns, such as the glaciated shield regions of Canada and northern Europe, a *deranged* pattern (g) is in evidence, with no clear geometry. This will be seen in the Jackson, Michigan, topographic map in Lab Exercise 16, ✱ SECTION 4.

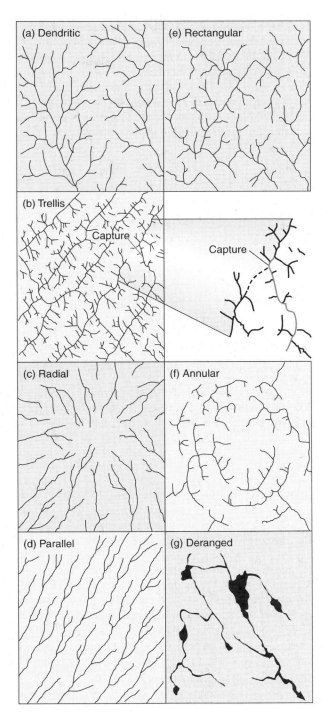

Figure 15.2
Seven most common drainage patterns
[After A. D. Howard, "Drainage analysis in geological interpretation: A summation," *Bulletin of American Association of Petroleum Geologists* 51 (1967), p. 2248. Adapted by permission.]

The low-lying area near a stream channel that is subjected to recurrent flooding is a **floodplain**, formed when the river leaves its channel during times of high flow. Thus, when the river channel changes course or when floods occur, the floodplain is inundated with water. When the water recedes, alluvial deposits generally mask the underlying rock. Episodic flood events are often devastating as we witnessed during the Midwest floods of summer 1993; Ohio River floods of late winter 1996–1997; and Red River floods that drowned Fargo and Grand Forks, North Dakota, and the regions surrounding Winnipeg, Manitoba, in spring 1997; or North Carolina in 1999 following the landfall of three hurricanes. Figure 15.3 illustrates a characteristic floodplain, with the present river channel embedded in the plain's alluvial deposits.

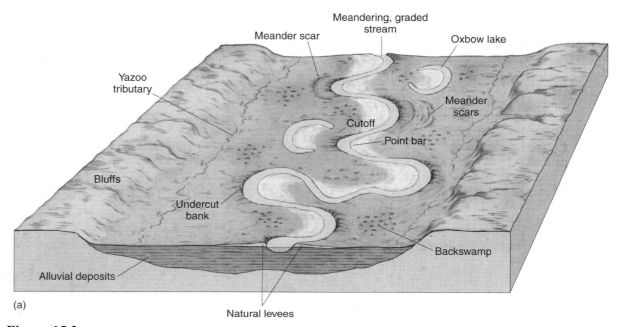

Figure 15.3
Typical floodplain landscape and related geomorphic features

On either bank of most streams, **natural levees** develop as by-products of flooding. When flood waters arrive, the river overflows its banks, loses velocity as it spreads out and drops a portion of its sediment load to form the levees. Notice on Figure 15.3 an area labeled **backswamp** and a stream called a **yazoo tributary**. The natural levees and elevated channel of the river prevent this tributary from joining the main channel, so it flows parallel to the river and through the backswamp area. When a meandering stream erodes its outside bank as the curve migrates downstream, the neck of land created by the looping meander eventually erodes through and forms a *cutoff*. When the former meander becomes isolated from the rest of the river, the resulting **oxbow lake** may gradually fill in with silt or may again become part of the river when it floods.

Philipp, Mississippi, Quadrangle and Photo Stereopair

Topographic Map #2 is a portion of the Philipp, Mississippi, 7.5-minute quadrangle in northwestern Mississippi. The Tallahatchie River flows through the landscape from west to east. (Remember to use the topographic map symbol legend inside the front cover of this manual.)

Analyze **Topographic Map #2** and the photo stereopair for this same area closely and answer the following questions and completion items (Figure 15.4).

1. What is the contour interval on this map segment? _____ Given this value, determine the maximum

    relief shown: _____ What is the highest spot elevation ("X") on the map

    and in what PLSS (Public Land Survey Section) section is it located? _____

    _____

2. How are the levees along the river portrayed by the elevation contours? How high are these levees (the river to the levee top)? Note the contours along Pecan Point near the well. How well do these show up on the

    photo stereopair? Can you identify them? _____

    _____

    _____

3. What is the water body called next to Jones Chapel and how did it form? Can you see this in the photo stereopair? Is the cemetery visible? Use the illustration in Figure 15.3 to identify this water feature.

    _____

    _____

4. How many similar such features, identified in question #3, can you find? _____

    What is the name of the largest one of these water features on the map? _____

5. In the Philipp quad, how many meander cutoffs can you identify? _____
    Are there distinct meander patterns (scars on the landscape once occupied by the meandering river)?

    Describe their appearance; how do you identify them? _____

6. Do the fluvial features have any effect on road patterns? Explain and give some examples. _____

    _____

    _____

7. Are there any backswamp areas identified (see Figure 15.3 for an illustration)? _____ In what sections

    of the map are the backswamp areas? _____

How did you identify these as backswamp areas on the map? Explain. _____

_____

8. What type of landform is Pecan Point? _____

_____

9. Relative to wetlands, swamps, and flooded areas, describe your observations when you compare and contrast

the topographic map and the stereo photos. _____

_____

_____

10. The National Aerial Photography Program (NAPP) scenes are at a 1:40,000 scale (smaller scale) whereas the topographic map is at a scale of 1:24,000 (larger scale). Are there details that you can see on the larger scale map that you cannot see on the smaller scale imagery? Describe the differences between the two figures.

_____

_____

_____

11. Optional Google Earth™ activity, Drainage patterns. For the KMZ file and questions go to mygeoscience-place.com. Then, click on the cover of *Applied Physical Geography: Geosystems in the Laboratory.*

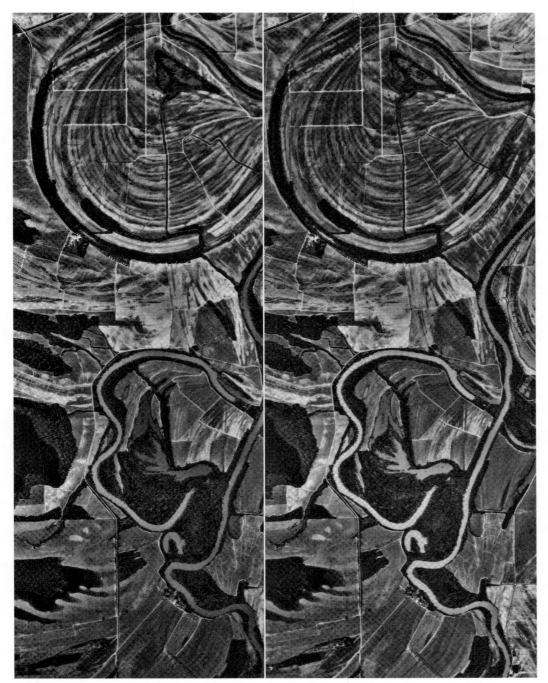

Figure 15.4
Philipp, Mississippi, photo stereopair. North is to the right, west to the top. (NAPP, USGS)

Omaha North Quadrangle

Streams often are used as natural political boundaries, but it is easy to see how disagreements might arise when boundaries are based on river channels that shift around. For example, the Ohio, Missouri, and Mississippi rivers can shift their positions quite rapidly during times of flood, thus creating a mismatch between the boundary and river. Carter Lake, Iowa, provides us with a case in point as shown in **Topographic Map #3**. The Nebraska-Iowa border was originally placed mid-channel in the Missouri River. In 1877, the meander loop that curved around Carter Lake, Iowa, was cut off by the river, thus leaving the town surrounded by Nebraska. The new oxbow lake is called Carter Lake. Today, the state boundary line still follows the former channel meander placing this Iowa enclave clearly within Nebraska!

Topographic Map #3 is a portion of the Omaha North 7.5-minute quadrangle, in extreme eastern Nebraska. The Missouri River flows through the landscape from north to south.

Analyze **Topographic Map #3** and inset map and answer the following completion items.

1. What is the contour interval on this map segment? _____

   Given this value, determine the relief in Carter Lake. _____
   What are the elevations of several bench marks ("BM") that surround the Carter Lake area (N, NE, SE, and SW)?

   _____

2. Estimate the area in square miles of Carter Lake (area within the oxbow lake). (To estimate, establish boxes or lay a grid over Carter Lake using the graphic scale, multiply two sides to determine the area in each grid

   box, then add them together.) _____

   Show your work: _____

   in square kilometers (conversions on lab manual back flap). _____

3. Using the graphic scale on the map, how wide (in feet) is the neck of Iowa state land along Abbott Drive?

   _____

4. The town of Carter Lake is in portions of which land survey sections? _____

5. In your opinion, what would happen in a flood of the Missouri River? How high would the river have to rise

   to connect with its former channel and Carter Lake? _____

   _____

   _____

   _____

6. Do you identify any industrial activity in Carter Lake? Describe. (Refer to map symbols on the inside front

   cover of this manual.) _____

   _____

   _____

7. Since the last revision of this map a new levee has been built closer to the river. How do you think this narrowing of the river channel and restricted floodplain will affect future flood events? Explain.

_____

_____

_____

8. Are there any closed *depression contours* (contour lines designating depressions) in Carter Lake? What objects do they surround? Analyze and explain.

_____

_____

9. Optional Google Earth™ activity, Sacramento River and political boundaries. For the KMZ file and questions go to mygeoscienceplace.com. Then, click on the cover of *Applied Physical Geography: Geosystems in the Laboratory.*

✳ SECTION 3

Cumberland Quadrangle and Photo Stereopair

In folded landscapes, ridges and valleys are the prominent *topographic features*, reflecting the **anticlines** and **synclines** which are the *structural landscape controls*. Anticlines are the upfolds and synclines are the downfolds. In the Ridge and Valley geomorphic province of the Appalachians, **water gaps** occur where a river cuts through an *anticlinal ridge*. Such a geomorphic feature is thought to form when a preexisting stream course flows across a landscape with buried geologic structures. As the overburden cover is removed, the downcutting stream cuts into the underlying structure. The stream valley is described as *superposed* where Wills Creek cuts through an anticlinal ridge (Haystack Mountain and Wills Mountain) in this example from Cumberland, Maryland. The importance of these water gaps to early transportation and migration is obvious. Note the abandoned canal in the southeast portion of the map.

Wind gaps are col-like passes that were formerly occupied by a stream that subsequently was captured and incorporated into another stream drainage. These abandoned stream valleys are generally dry. Two remarkable examples of these features are on the portion of the Cumberland 7.5-minute quadrangle presented as **Topographic Map #4.**

Figure 15.5 shows two views (a contour map and a stereoscopic contour map) of a hypothetical anticline and syncline landscape. As you did in Lab Exercise 14, use stereolenses to view Figure 15.5b in three dimensions and note selected features marked on the contour map. The anticlines (A) are separated by synclines (B), and the heavy lines (C) and (D) mark the axes along which the folding occurred forming the ridges and troughs. The arrows at (E) point downhill, away from the crest of the anticline, and the arrows at (F) also point downhill, towards the trough of the syncline.

Refer back to ✳ SECTION 1 and the review of drainage basins and drainage patterns. Viewing the stereoscopic contour map, it is easy to imagine the direction in which water would flow down the anticlines into streams that will occupy the synclines, establishing the *trellis* drainage pattern in these ridge and valley regions. Keep this stereoscopic image and the trellis drainage pattern in mind when analyzing Topographic Map #4.

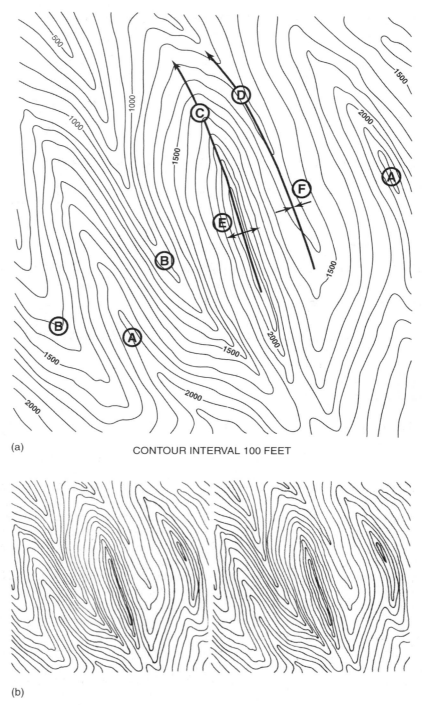

(a)

CONTOUR INTERVAL 100 FEET

(b)

Figure 15.5
Contour map (a) and stereoscopic contour map (b) of the same anticline and syncline landscape. (From Horace MacMahan, Jr., *Stereogram Book of Contours*. Copyright © 1995, by Hubbard Scientific Company, pp. 24–25. Reprinted by permission of American Educational Products, Hubbard Scientific.)

Analyze **Topographic Map #4** and the set of stereo photos of the same area in Figure 15.6 to answer the following questions and completion items.

1. What is the contour interval on this map segment? _____ What is the scale of this map? _____

   Convert this to a written scale. Show your work. _____

   _____

   _____

2. Wills Creek at The Narrows is a *superposed stream* forming a *water gap*. How wide is the water gap gorge (distance between the 1560-foot contour to the north and the 1480-foot contour to the south)?

   _____

   How deep is the water gap at "The Narrows"? _____. Examine this feature in the photo

   stereopair. Do you discern any floodplain or river terraces in the water gap? _____

   _____

   _____

3. What is the relief between the floor of the wind gap (Braddock Road) and the floor of the valley to the north

   in which Braddock Run flows? _____ A wind gap was formed when the lower course of Braddock

   Run was diverted (captured) by a tributary of Wills Creek (stream in the water gap). How wide is the wind

   gap, measured from the 1400' contour line on the north rim to the 1500' contour line on the south rim? _____

   How deep is the gap, measured from the water tank on the north rim to Braddock Road in the gap? _____

4. How much higher in elevation was the valley in which Braddock Run flows at the time the stream that cut

   the gap was captured, that is, when the valley and the floor of the wind gap were at the same elevation?

   _____

   Explain your thinking on this question. _____

   _____

   _____

   _____

5. Imagine a topographic profile from west of Narrows Park, along Braddock Run at the 700-foot contour line,

   to the 760-foot contour line south of Allegany High School at the south-east corner of the cemetery. How

   long is this profile line in straight distance? _____

What is the highest point along the profile? _____

What is the relief from Allegany High School to the highest point along the profile? (south of the water gap):

_____

_____

6. What topographic features appear to be controlling the stream course of the North Branch of the Potomac River (joins Wills Creek along the southeast portion of the map) and what are the names of the structural landscape controls? _____

_____

_____

7. How many crossings of Wills Creek do you count in the stereo photo? In the topographic map? _____

_____

_____

8. Optional Google Earth™ activity, Cumberland Gap. For the KMZ file and questions go to mygeoscience-place.com. Then, click on the cover of *Applied Physical Geography: Geosystems in the Laboratory.*

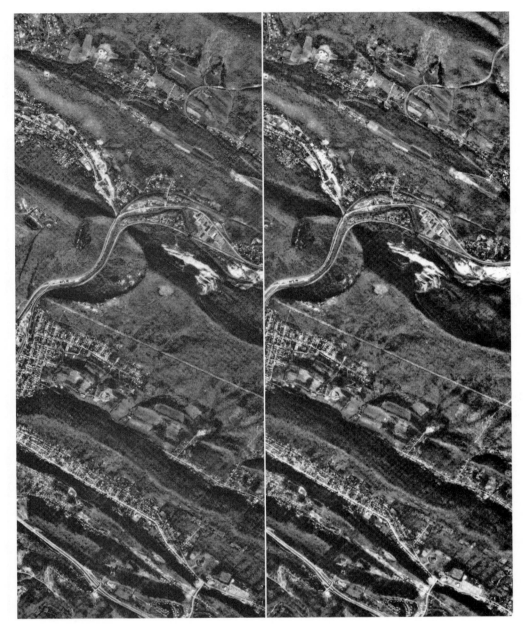

Figure 15.6
Cumberland, Maryland, photo stereopair. North is to the right, west to the top. (NAPP, USGS)

Name: _____

Date: _____

Laboratory Section: _____

Score/Grade: _____

Lab Exercise 16

Topographic Analysis: Glacial Geomorphology

About 77% of Earth's freshwater is frozen, with the bulk of that ice sitting restlessly in just two places—Greenland and Antarctica. The remaining ice covers various mountains and fills some alpine valleys. More than 32.7 million cubic kilometers (7.8 million cubic miles) of water is tied up as ice.

A **glacier** is a large mass of ice, resting on land or floating as an ice shelf in the sea adjacent to land. Glaciers are not frozen lakes or groundwater ice. Instead, they form by the continual accumulation of snow that recrystallizes under its own weight into an ice mass.

Glacial erosion and deposition produce distinctive landforms that differ greatly from the way the land looked before the ice came and went. Worldwide, glacial ice is in retreat, melting at rates exceeding anything in the ice record. In the European Alps alone, some 75% of the glaciers have receded in the past 50 years, losing more than 50% of their ice mass since 1850. At this rate, the European Alps will have only 20% of their preindustrial glacial ice left by 2050.

In this lab exercise topographic map interpretation continues. Specific exogenic geomorphic processes and the resulting landforms from glacial (moving ice) action are analyzed using topographic and stereoscopic contour maps, a photo stereopair, and illustrations. A complete legend to topographic map symbols is inside the front cover of this lab manual. Lab Exercise 16 features four sections and one Google Earth™ activity.

Key Terms and Concepts

ablation
alpine glacier
cirque
continental glacier
drumlin
equilibrium line
esker
glacier

kame
kettle
moraine
outwash plain
tarn
till plains
valley glacier

Objectives

After completion of this lab, you should be able to:

1. *Analyze* and *describe* geomorphic processes using topographic maps and Google Earth™ imagery.
2. *Use* data to *graph, calculate,* and *analyze* glacial mass balance.
3. *Interpret* and *describe* several glacial processes and the characteristic landscapes that result, including erosional and depositional features of alpine and continental glaciers.

Materials/Sources Needed

pencil
ruler
stereolenses

string
calculator

Lab Exercise and Activities

✳ SECTION 1

Glaciated Landscapes

Begin by reviewing relevant sections in a physical geography text, or *Geosystems*, that cover glacial processes and landscapes. Here are a few essentials. A **glacier** is a large mass of perennial (year-round) ice, resting on land or floating shelflike in the sea adjacent to land. Glaciers are not frozen lakes or groundwater ice but form by the continual accumulation of snow that recrystallizes into an ice mass. They move under the pressure of their own great mass and the pull of gravity. With few exceptions, a glacier in a mountain range is called an **alpine glacier**, or *mountain glacier*. The name comes from the Alps where such glaciers abound. On a larger scale than individual alpine glaciers, a continuous mass of ice is known as a **continental glacier** and in its most extensive form is called an *ice sheet*. Two additional types of continuous ice cover associated with mountain locations are designated as *ice caps* and *ice fields*.

Alpine glaciers form in several subtypes. One prominent type is a **valley glacier**, an ice mass constricted within a valley that was originally formed by stream action. As a valley glacier flows slowly

downhill, the mountains, canyons, and river valleys beneath its mass are profoundly altered by its passage. Most alpine glaciers originate in a mountain snowfield that is confined in a bowl-shaped recess. This scooped-out erosional land form at the head of a valley is called a **cirque**. In the cirques where the valley glaciers originated, small mountain lakes, called tarns, may form. Some cirques may contain small, circular, stair-stepped lakes, called **paternoster** ("our father") lakes for their resemblance to rosary (religious) beads. As the cirque walls erode away, sharp ridges form, dividing adjacent cirque basins. These **arêtes** ("knife edge" in French) become the sawtooth, serrated ridges in glaciated mountains. Two eroding cirques may reduce an arête to a saddle-like depression or pass, called a **col**. A **horn** (pyramidal peak) results when several cirque glaciers gouge an individual mountain summit from all sides. Figure 16.1 shows a cross section of a typical retreating alpine glacier.

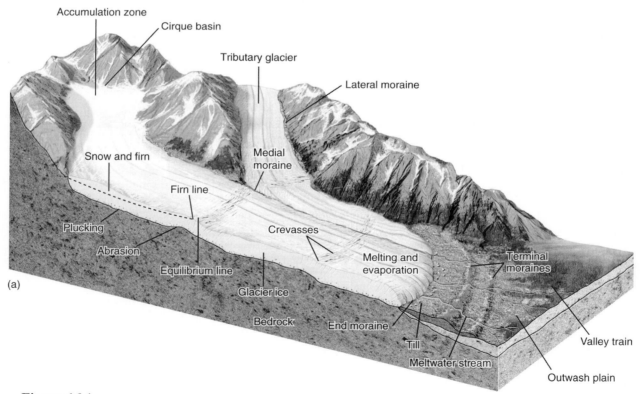

Figure 16.1
Cross-section of a typical retreating alpine glacier

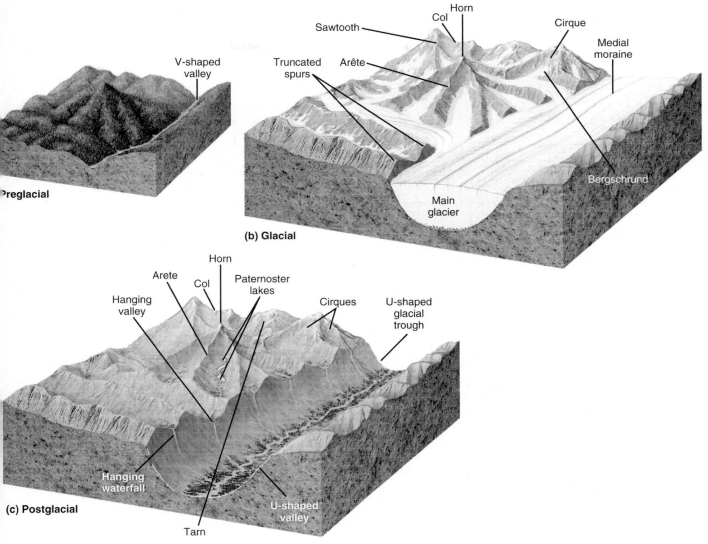

Figure 16.2
The work of alpine glaciers, during and after glaciation. (After W.M. Davis.)

Mount Darwin, California Quadrangle

Analyze **Topographic Map #5**. The topographic map was made in 1994. The NAPP photos were taken in 1999.

1. A cirque basin may serve as an accumulation zone for alpine glaciers. Using contour lines on the mountains and on the glaciers, how many ice-filled cirques (scooped out amphitheater-like basins) do you identify on

   the map? _____ What is the elevation of the highest glacial ice surface on the map?

   _____

2. What type of glacial feature is the unnamed peak, with an elevation of 13,253', north of Mount Darwin?

   _____

   _____

3. What is the relief (elevation difference) between Goethe Peak and Goethe Lake? _____

   _____

4. a) What is the linear distance between Goethe Peak and the south shoreline of Goethe Lake? _____

_____

b) In terms of slope, state this distance in <u>feet per mile</u>: _____

c) Take the <u>relief in feet</u> that you determined in question 2 and <u>divide it by the distance between the two locations in feet</u> (a), to determine the *percent grade* (<u>relief divided by distance</u>). What percent grade did you

determine? (Show your work.) _____

_____

5. Locate Sky High Lake in section 16 on the topo map. How was this lake formed? _____

_____

_____

6. What is the term for the string of lakes in Darwin Canyon?

_____

7. South of Goethe Lake find Mount Goethe topographic map #5 and photo stereopairs Figure 16.3b.

a) How many cirque glaciers do you count that surround Mount Goethe? _____

b) How many are on the north side? _____ South side? _____

c) Describe the orientation of the slopes with snow on them on the topographic map and the photo stereopair. Is there more snow on north facing slopes or on south facing slopes? Why would there be more snow on one slope than another? How would this affect the amount of glaciation that would occur on each slope?

_____

_____

_____

d) After looking at the photo stereopair, describe the steepness of the north and south facing slopes of Mount Goethe and Mount Lamark. Which side has steeper slopes? Why do you suppose this is?

_____

_____

_____

8. a) What term would you use to describe the ridge west of Mount Goethe? _____

_____

b) What is the term for the feature found on the ridge between Mount Goethe and Muriel Peak?

_____

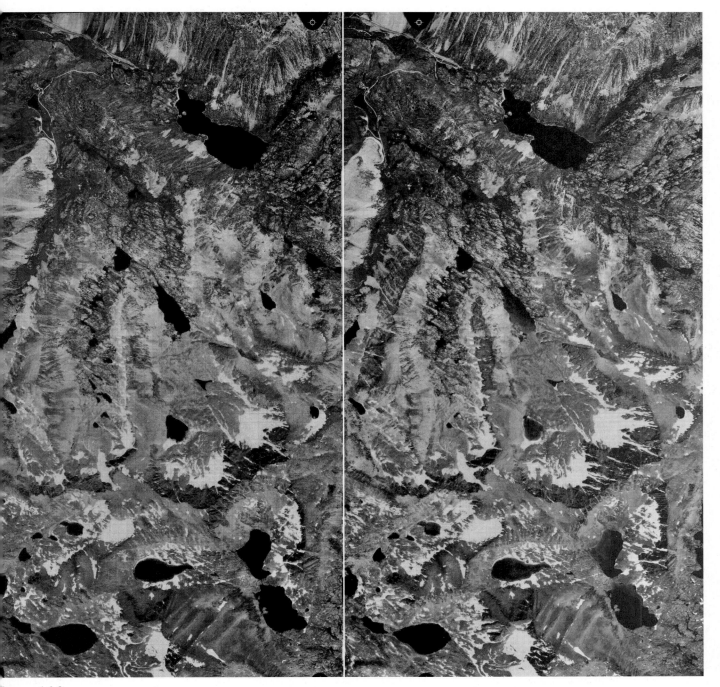

Figure 16.3
Photo stereopair of the Mount Darwin area of California. In this arrangement north is at the top.

9. Optional Google Earth™ activity, Mount Darwin, CA. For the KMZ file and questions go to mygeoscience-place.com. Then, click on the cover of *Applied Physical Geography: Geosystems in the Laboratory.*

✳ SECTION 2

Glacial Mass Balance

A glacier is fed by snowfall and other moisture sources and is wasted by losses from its upper and lower surfaces and along its margins. The combined effect of these losses is called **ablation**. A glacier's area of accumulation is, logically, at colder higher elevations. The zone where accumulation gain balances ablation loss is the **equilibrium line**—Figure 16.1. Glaciers achieve a positive net balance of mass—*grow larger*—during colder periods with adequate precipitation. In warmer times, the equilibrium line migrates to a higher elevation and the glacier retreats—*grows smaller*—due to its negative net balance. Internally, gravity continues to move a glacier forward even though its lower terminus is in retreat.

Activity with the mass balance of a glacier

1. Using the data in Table 16.1, note the net mass balance for the South Cascade Glacier, Washington, for 1959 through 2009. Net mass balance is specified as centimeters of water equivalent spread over the entire glacier. Plot these positive and negative values with bar graphs on Figure 16.4 and determine the absolute sum of the net balance values. The glacier has gone through a net wastage between 1959 and 2009 at lower and middle elevations. In 2005, the glacier experienced its second greatest ice wastage. The 2003–2005 combined net balance loss of 6.20 meters is the largest three-year loss in the four decades the glacier has been studied.

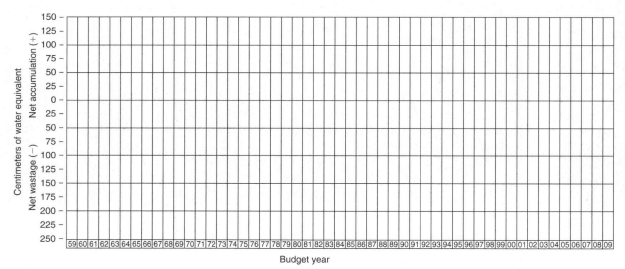

Figure 16.4
South Cascade Glacier net mass balance graph.

2. When did the trend in negative balances begin to dominate the South Cascade Glacier's mass balance?

Explain. _____

_____

Table 16.1
South Cascade Glacier Net Mass Balance Data (cm) 1959 to 2009*

YEAR	Net Balance	YEAR	Net Balance	YEAR	Net Balance
1959	70	1976	95	1993	−123
1960	−50	1977	−130	1994	−160
1961	−110	1978	−38	1995	−69
1962	20	1979	−156	1996	10
1963	−130	1980	−102	1997	63
1964	120	1981	−84	1998	−186
1965	−17	1982	8	1999	102
1966	−103	1983	−77	2000	38
1967	−63	1984	12	2001	−157
1968	1	1985	−120	2002	55
1969	−73	1986	−61	2003	−210
1970	−120	1987	−206	2004	−165
1971	60	1988	−134	2005	−245
1972	143	1989	−91	2006	−159
1973	−104	1990	−11	2007	−20
1974	102	1991	7	2008	−30
1975	−5	1992	−201	2009	−20

*Data from R. M. Krimmel, *Water, Ice, Meteorological, and Speed Measurements at South Cascade Glacier, Washington, 2002 Balance Year.* USGS Water Resources Report, Tacoma, Washington, 2002; http://ak.water.usgs.gov/glaciology/all_bmg/3glacier_balance.txt; Bidlake, W.R., Josberger, E.G., and Savoca, M.E., 2007, *Water, Ice, and Meteorological Measurements at South Cascade Glacier, Washington, Balance Years 2004 and 2005:* U.S. Geological Survey Scientific Investigations Report 2007-5055; and personal communication.

Figure 16.5
South Cascade Glacier photo comparison 1928–2000. In 1928, the glacier's volume in water equivalent was 0.32 km^3 (0.077 mi^3), but by 2001, it had decreased to 0.16 km^3 (0.038 mi^3).

3. What is the absolute sum of the net balance amounts for 1959–2009? _____

4. What local, regional, and global conditions might produce these trends in the mass balance of the South

   Cascade Glacier? _____

   _____

   _____

   _____

Alpine Glacier and Mount Rainier Quadrangle

An alpine (valley) glacier has profound effects on the landscape. Figure 16.2 illustrates views during and after a glacier's passage. A typical river valley has a characteristic **V**-shape and stream-cut tributary valleys. During active glaciation, glacial erosion and transport actively remove much of the regolith (weathered bedrock) and the soils that covered the stream valley landscape. When climates warm and the ice has retreated, the valleys are exposed as **U**-shaped glacially carved valleys, greatly changed from their previous stream-cut form. You can see the oversteepened sides and the straightened course of the valley.

Figure 16.6a shows the contour lines for an alpine glacier scene with key portions labeled, whereas Figure 16.6b is a stereocontour map of an alpine glacier area to view with stereolenses. Please compare this to the alpine glaciation illustration in Figure 16.2. Mount Rainier is featured in **Topographic Map #6**— a dormant composite volcano in the Cascade Range that is covered by several glaciers and glacial features.

Activities and completion items related to alpine glacier geomorphology

1. Figure 16.6 shows a contour map and a stereoscopic map of a hypothetical landscape that has been shaped by alpine glaciation. As glaciers carved their way down preexisting V-shaped stream valleys, they widened the valleys into U-shaped valleys separated by sharp ridges. Hanging valleys/troughs left stranded above the main valley floor will be occupied by streams that form picturesque waterfalls as they plunge over the steep edge.

   Glaciers may scour out the bedrock of the valley floor, leaving depressions that, as the glacier retreats and a stream reoccupies the valley, subsequently fills with water, forming a glacial lake called a **tarn**. Steep-walled cirques mark the origin of the glaciers high on the mountain slope. Keep this stereoscopic contours map, along with the stereo pairs and Mount Darwin topo map, in mind when analyzing the topographic map of Mount Rainier.

After reviewing Figures 16.1 and 16.2, answer the following questions about Figure 16.6a:

If you wanted to go swimming in a tarn which letters would you look for? _____

If you were a rock climber and wanted to scale the steep sides of the valley which letters would you look

for? _____

Which letters indicate the location of a pass on the ridge between two valleys? _____

If you wanted to climb the horn created by three glaciers which letter would you look for? _____

What is the elevation of the peak? _____

Find and label the following features on Figure 16.6a: a hanging valley; the U-shaped valley formed by the main glacier; a cirque; a horn; an arête (the narrow ridge that separates two valleys); a col (a pass on an arête); and a chain of paternoster lakes (if they were filled with water).

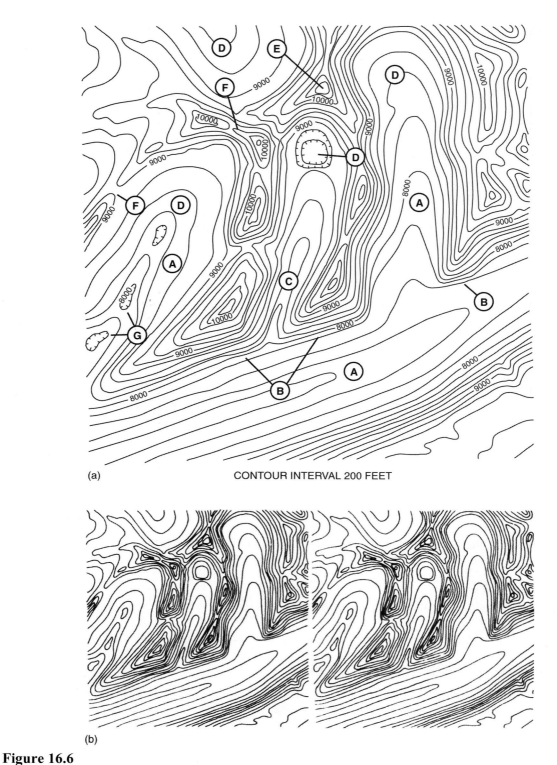

(a)

CONTOUR INTERVAL 200 FEET

(b)

Figure 16.6
Alpine glacier landscape. Contour map (a) and stereoscopic contour map (b) of the same alpine glacial landscape. (Adapted from Horace MacMahan, Jr., *Stereogram Book of Contours*. Copyright © 1995, by Hubbard Scientific Company, pp. 12–13. Reprinted by permission of American Educational Products—Hubbard Scientific.)

2. Refer to Mount Rainier National Park topographic quadrangle—**Topographic Map #6**. What is the scale of

    this map? _____

    Please convert this to a written scale. Show your work. _____

    What is the contour interval of the topo map? _____.

3. Name several glaciers on which you identify medial and lateral moraines.

    _____

    _____

    _____

4. What is the named cirque on the topographic map? _____

    _____

5. What type of feature is labeled Nisqually Cleaver? List several other features that are the same type, including

    at least two that are not named Cleaver. _____

    _____

    _____

    _____

6. The summit of Mount Rainier is at 14,410 ft (4392 m). What is the lowest elevation shown on this map?

    Where is this location? Give its name. _____

    Therefore, the relief on this map segment is _____

    What is the linear distance between these high and low points? _____

    In terms of slope, state this distance and relief in *feet per mile*: _____

    State this in terms of *percent grade* along an ideal slope between these two points (relief in ÷ distance

    in feet): _____ %.

    Show your work: _____

    _____

Continental Glaciation and the Jackson, Michigan, Topographic Quadrangle

Continental glaciers advanced and retreated over North America and Europe producing many erosional and depositional features. Figure 16.7 illustrates some of the most common depositional features associated with the passage of a continental glacier. The unsorted and unstratified deposits of gravel, sand, and clay form **moraines**, including ground and terminal moraines. Many relatively flat plains of *unsorted* coarse till are formed behind terminal moraines. These **till plains** typically have low, rolling relief, deranged drainage patterns, and the following depositional features. **Drumlins**, smooth hills made of till shaped by the ice, are oriented in the direction of the glacier's movement. **Eskers** are curving, narrow deposits of coarse gravel left by meltwater stream deposits in tunnels beneath the ice. Often eskers end in a delta. **Kames** are small hills of poorly sorted sand and gravel that collected in depressions in the surface of a glacier.

Beyond the morainal deposits, glacio-fluvial **outwash plains** of *stratified drift* feature stream channels that are meltwater-fed, braided, and overloaded with debris deposited across the landscape. **Kettles** are depressions left by melting blocks of ice that were buried in the drift and are found in both ground moraines and in outwash plains.

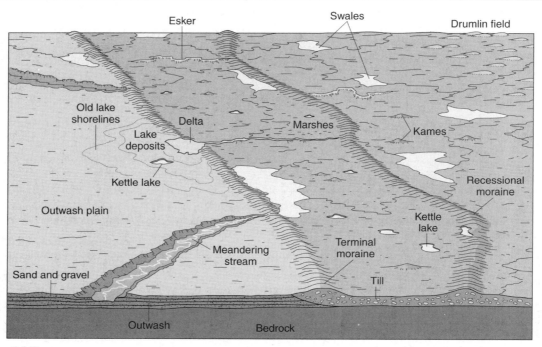

Figure 16.7
Common depositional landforms produced by glaciers

Activities and completion items related to continental glaciation

1. Figure 15.8 shows two views (a contour map and a stereoscopic contour map) of a hypothetical landscape that has been shaped by continental glaciers. In Figure 16.8a, note the unsorted and unstratified deposits of gravel, sand, and clay that form **moraines**, including terminal moraines (A) and interlobate moraines (B). Use stereo lenses to view the typical continental features in Figure 16.8b in three dimensions.

After reviewing Figures 16.7 and 16.8, answer the following questions about Figure 16.8a:

If you were looking for drumlins which letters would you look for? _____

Which letter shows kames? _____

If you wanted to walk along the crest of an esker which letter would you look for? _____

If you were looking for kettles which letters would you look for? _____

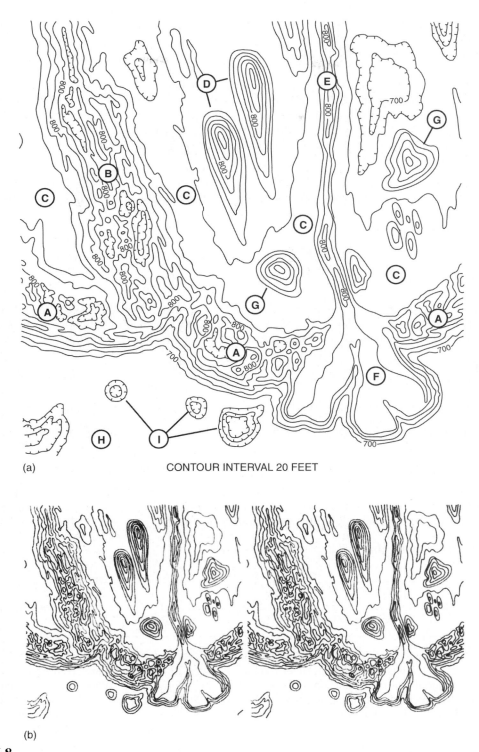

(a)

CONTOUR INTERVAL 20 FEET

(b)

Figure 16.8
Continental glaciated landscape. Contour map (a) and stereoscopic contour map (b) of the same continental glaciated landscape.
(From Horace MacMahan, Jr., *Stereogram Book of Contours*. Copyright © 1995, by Hubbard Scientific Company, pp. 14–15. Reprinted by permission of American Educational Products, Hubbard Scientific.)

Which letter is located on the ground moraine? _____

Which letter is located on the outwash plain? _____

Once again refer to Figure 15.2g in Lab Exercise 15 or Figure 14.9 in *Geosystems* or Figure 11.9 in *Elemental Geosystems* and the review of drainage basins and drainage patterns. Viewing the stereoscopic contour map, imagine the deranged drainage patterns that will result from the disruption of former surface patterns by the erosional and depositional work of continental glaciation.

Keep Figures 16.7 and 16.8 in mind when analyzing **Topographic Map #7** of the formerly glaciated area around Jackson, Michigan.

2. Refer to the Jackson, Michigan, topographic quadrangle—**Topographic Map #7**. What is the scale of this

map? _____

Convert this to a verbal scale. Show your work. _____

What is the contour interval? _____

3. A sinuously curving, narrow ridge of coarse sand and gravel is called an **esker**. Eskers form along the channel of a meltwater stream that flows beneath a glacier, in an ice tunnel, or between ice walls beneath the glacier. As a glacier retreats, the steep-sided esker is left behind in a pattern roughly parallel to the path of the glacier.

Locate and identify by name a prominent esker on the map: _____

What is the average elevation of this esker? _____

How long is this esker in miles? (Use a string placed along the esker, then pull it straight and compare it to

the scale.) _____

4. Identify the drainage pattern (using Figure 14.2 or Figure 14.9 in *Geosystems* or Figure 11.9 in *Elemental Geosystems*), for the portion of the topo map that is generally covered by the Kalamazoo moraine (southwest part of map). Explain what evidence you used to determine this.

5. Is this area well or poorly drained? Explain your answer.

_____

_____

_____

6. Sometimes an isolated block of ice, perhaps more than a kilometer across, persists in a *ground moraine*, an outwash plain, or valley floor after a glacier retreats. Perhaps 20 to 30 years are required for it to melt. In the interim, material continues to accumulate around the melting ice block. When the block finally melts, it leaves behind a steep-sided hole. Such a feature then frequently fills with water. This is called a **kettle**. Locate several named kettles that are ponds or lakes on this map and identify their general location:

_____

_____

7. What is the highest elevation on the map segment? Location and name: _____

What is the lowest elevation on the map segment? Location and name: _____

Name: _____   Laboratory Section:_____

Date: _____   Score/Grade: _____

Lab Exercise 17

Topographic Analysis: Coastal and Arid Geomorphology

In addition to streams of water and ice—rivers and glaciers—carving their way across Earth's surface, other exogenic processes are also at work. Waves are relentless in their efforts to reshape the contours of the continents, whether sandy **beaches** or rocky cliffs, along the narrow contact zone between the landmasses and oceans. Anthropogenic (human) processes are also very important, as we will examine in the development of the New Jersey coast. And landforms that are characteristic of arid climates are limited to those global areas where adequate precipitation is always lacking, allowing wind and streams to remodel the surface.

Topographic map interpretation continues in this lab exercise. Specific geomorphic processes and the resulting landforms from coastal (wave) action and the work of streams in arid climates are analyzed using topographic maps, photo stereopairs, and illustrations. Once again refer to the complete legend of topographic map symbols located on the inside front cover of this lab manual. Lab Exercise 17 features five sections and four Google Earth™ questions.

Key Terms and Concepts

alluvial fan
barrier spit
bay barrier
beach

lagoon
littoral zone
tombolo
wave-cut platform

Objectives

After completion of this lab, you should be able to:

1. *Analyze* and *describe* geomorphic processes using topographic maps and Google Earth™ imagery.
2. *Interpret* and *describe* several coastal processes and the characteristic landscapes that result, including erosional and depositional features.
3. *Interpret* and *analyze* patterns of development in sensitive coastal areas, and *suggest* appropriate development strategies.
4. *Interpret* and *describe* several arid climate processes and the characteristic landscapes that are produced, including alluvial fans, buttes, and mesas.
5. *Analyze* and *evaluate* the water budget of an exotic stream, and *predict* future water issues.

Materials/Sources Needed

pencil
ruler
calculator
stereolenses

Lab Exercise and Activities

✳ SECTION 1

Coastal Features and the Point Reyes Quadrangle and Stereophotos

Begin by reviewing relevant sections in a physical geography text, or *Geosystems*, that covers coastal processes and landscapes. Here are a few essentials. Most of Earth's coastlines are relatively new and are the setting for continuous change. The land, ocean, atmosphere, Sun, and Moon interact to produce tides, currents, and waves that create erosional and depositional features along the continental margins. A dynamic equilibrium among the energy of waves, wind, and currents; the supply of materials; the slope of the coastal terrain; and the fluctuation of relative sea level, produces coastline features of infinite variety. The coastal environment is called the **littoral zone**. (Littoral comes from the Latin word *litus* meaning "seashore.") The littoral zone spans both land and water. Landward, it extends to the highest water line that occurs on shore during a storm. Seaward, it extends to the point at which storm waves can no longer move sediments on the seafloor (usually at depths of approximately 60 m or 200 ft). The specific

contact line between the sea and the land is the *shoreline*, and adjacent land is considered the *coast*.

The active margins of the Pacific along the North and South American continents are characteristic coastlines affected by erosional landform processes. Erosional coastlines tend to be rugged, of high relief, and tectonically active, as expected from their association with the leading edge of drifting lithospheric plates (see plate tectonics discussion in Lab Exercise 12).

Figure 17.1 presents features commonly observed along an erosional coast. *Sea cliffs* are formed by the undercutting action of the sea. As indentations are produced at water level, such a cliff becomes notched, leading to subsequent collapse and retreat of the cliff. Other erosional forms evolve along cliff-dominated coastlines, including *sea caves, sea arches,* and *sea stacks*. As erosion continues, arches may collapse, leaving isolated stacks in the water.

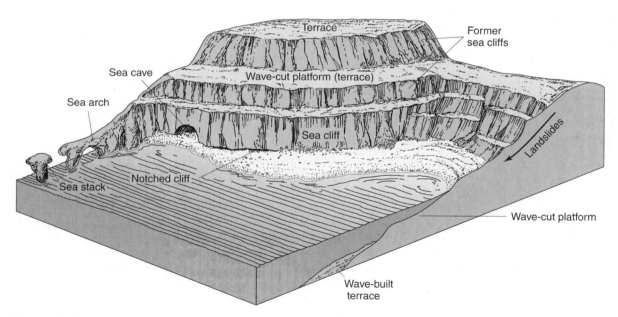

Figure 17.1
Characteristic coastal erosional landforms

Depositional coasts generally are located near onshore plains of gentle relief, where sediments are available from many sources, although such features can occur along all coasts. Characteristics of wave- and current-deposited landforms are illustrated in Figure 17.2 and may involve sediments of varying sizes. A **barrier spit** consists of material deposited in a long ridge, attached at one end, extending out from a coast; it partially crosses and blocks the mouth of a bay. A spit becomes a **bay barrier**, sometimes referred to as a *baymouth bar*, if it completely cuts off the bay from the ocean and forms an inland **lagoon**. Tidal flats and salt marshes are characteristic low relief features wherever tidal influence is greater than wave action.

A **tombolo** occurs when sediment deposits connect the shoreline with an offshore island or sea stack. A tombolo forms when sediments accumulate on a wave-built terrace that extends below the water.

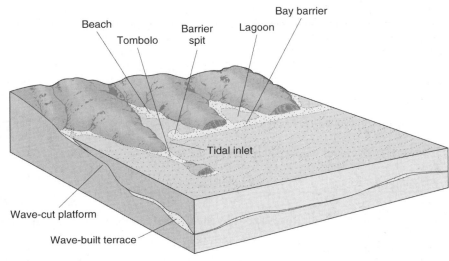

Figure 17.2
Characteristic coastal depositional landforms

Activities and completion items related to coastal geomorphology.

1. Refer to the Point Reyes topographic quadrangle—**Topographic Map #8**.

   What is the scale of this map? _____

   What is the contour interval? _____

2. Give the name and location of a barrier spit on the map segment.

   _____

3. List the names of one lagoon and two estuaries shown on the map. Note that "estero" means estuary in Spanish.

   _____

   _____

4. Using the topographic map symbol key on the inside front cover of your lab manual, what is the composition of

   Limantour Spit? _____

   _____

   _____

5. A **wave-cut platform**, or *wave-cut terrace*, is formed by wave action as a horizontal bench in the tidal zone. If the relationship between the land and sea level has changed over time, multiple platforms or terraces may arise like stairsteps back from the coast. These marine terraces are remarkable indicators of an emerging coastline. What evidence do you find of such a leveled terrace on the map? Describe. _____

_____

_____

6. Looking at the contour lines, can you determine the prevailing, effective wind direction along Point Reyes Beach? Explain. _____

_____

7. From Drakes Beach around to Chimney Rock there are examples of wave-cut cliffs and hanging valleys. Determine the height of several of these features above sea level. _____

_____

8. If you were in a small boat near Point Reyes and wanted to come ashore, would you consider docking near Chimney Rock? Explain why or why not. _____

_____

_____

9. In the photo stereopair in Figure 17.3, locate the lagoons and bays behind Limantour Spit (the *bay barrier* and *barrier spit*) that you saw on the topographic map. The topo map was prepared in 1954 and revised in 1971 whereas the photo pair was made in 1993–39 years later.

   a) Describe some specific differences you see in comparing the map and the photos. _____

   _____

   _____

   _____

   b) Are there any cultural features that appear on the photos and not on the map? Describe. _____

   _____

   _____

   _____

10. Optional Google Earth™ activity, Pt. Reyes, CA. For the KMZ file and questions, go to mygeoscience-place.com. Then, click on the cover of *Applied Physical Geography: Geosystems in the Laboratory*.

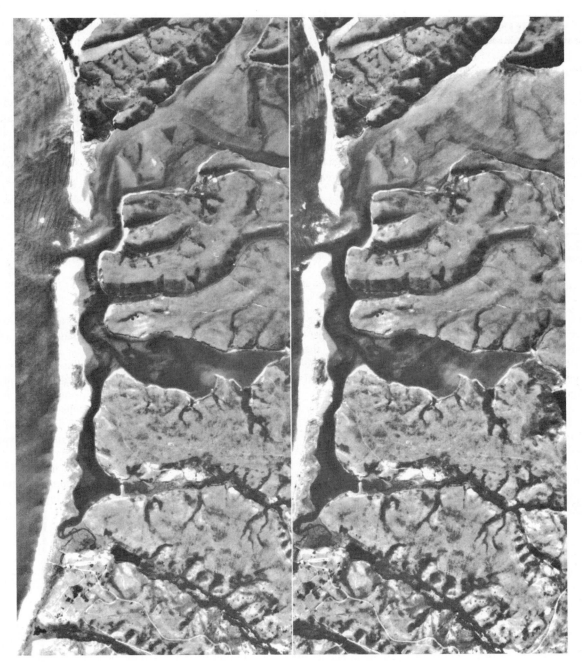

Figure 17.3
Photo stereopair of a portion of Point Reyes National Seashore, California. In this arrangement north is to the right, west to the top. (Photos by NAPP, USGS.)

✴ SECTION 2

Coastal Geomorphology and Land Use Planning

The coastal environment on the east coast of the United States has been heavily developed and extensively studied. Ian McHarg, in his book *Design with Nature*, proposed ecological planning principles to guide development of coastal resources. In this Google Earth™ exercise, you will use historic and current topographic maps, as well as Google Earth™ imagery, to analyze physical changes along the New Jersey coast, evaluate past planning decisions and the potential effects of global climate change. For the KMZ file and questions, go to mygeoscienceplace.com. Then, click on the cover of *Applied Physical Geography: Geosystems in the Laboratory*.

✴ SECTION 3

Ennis, Montana, 15-minute Quadrangle

Dry climates occupy about 26% of Earth's land surface and, if all semiarid climates are considered, perhaps as much as 35% of all land, constituting the largest single climatic region on Earth. Despite the general dryness, water remains the major erosional force in arid and semiarid regions.

In arid climates, with their intermittent water flow, a particularly noticeable fluvial landform is the **alluvial fan**, or *alluvial cone*, which occurs at the mouth of a canyon where it exits into a valley. The fan is produced by flowing water that loses velocity as it leaves the constricted channel of the canyon and therefore, drops layer upon layer of sediment along the base of the mountain block. Water then flows over the surface of the fan and produces a braided drainage pattern, shifting from channel to channel with each moisture event.

An interesting aspect of an alluvial fan is the natural sorting of materials by size. Near the mouth of the canyon, boulders and gravels are deposited, grading slowly to pebbles and finer gravels with distance out from the mouth. Then, sands and silts are deposited, with the finest clays and salts carried in suspension and solution all the way to the valley floor.

A remarkably symmetrical alluvial fan, produced by Cedar Creek, is on the Ennis, Montana, 15-minute quadrangle presented as **Topographic Map #9**.

Analysis and completion items for **Topographic Map #9**.

1. What is the contour interval on this map segment? _____

   What is the scale of this map? _____

2. What is the linear distance along the line stretching northwest (WNW) from the mouth of Cedar Creek Canyon (just east of Lawton Ranch, on the boundary between Sections 16 and 22) to the benchmark just west

   of the gravel pit (on the boundary between Sections 10 and 11), in miles? _____

   What is the relief between these two points? _____ In terms of slope, state this distance

   and relief in *feet per mile*: _____ State this in terms of *percent grade* along an ideal slope

   between these two points (relief in feet ÷ distance in feet): _____ % grade.

   Show your work: _____

   _____

3. The Cedar Creek alluvial fan covers approximately how many square miles? _____

Show your work (Use 1 mi^2 sections marked and numbered on the map to block out your

estimate.): _____

4. Entrenchment of a river into its own floodplain produces **alluvial terraces** on either side of the valley, which look like topographic steps above the river. What is the height of the alluvial terraces (local relief) south of

Shelhamer Ranch? _____

5. How would you characterize the Madison River channel? Describe. _____

_____

_____

6. Describe any human efforts to redirect flows of water on the topo map. _____

_____

_____

_____

7. Optional Google Earth™ activity, Cedar Creek Alluvial Fan in Ennis, MT. For the KMZ file and questions, go to mygeoscienceplace.com. Then, click on the cover of *Applied Physical Geography: Geosystems in the Laboratory.*

✳ SECTION 4

Mitten Buttes, Arizona, Quadrangle and Photo Stereopairs

Monument Valley straddles the Utah-Arizona border west of the Four Corners area. It is almost entirely within the Navajo Indian Reservation, and is accessed through the Navajo Tribal Park off U.S. 163. Part of this region is shown on **Topographic Map #10** and in the stereo photos in Figure 17.4.

The buttes, pinnacles, and mesas of arid landscapes are resistant horizontal rock strata that have eroded differentially. Removal of the less-resistant sandstone strata produces unusual desert sculptures—arches, windows, pedestals, and delicately balanced rocks. Specifi-

cally, the upper layers of sandstone along the top of an arch or butte are more resistant to weathering and protect the sandstone rock beneath.

The removal of all surrounding rock through differential weathering leaves enormous buttes as residuals on the landscape. If you imagine a line intersecting the tops of the Mitten Buttes shown in Figure 17.4, you can gain some idea of the quantity of material that has been removed. These buttes exceed 300 m (1000 ft) in height, similar to the Chrysler Building in New York City or First Canadian Place in Toronto.

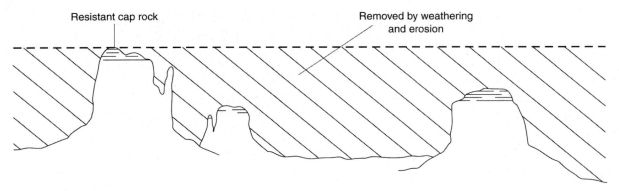

Figure 17.4
Rock removal by weathering, erosion, and transport in Monument Valley

Figure 17.5 is a photo stereopair of the buttes and part of a mesa in Monument Valley in **Topographic Map #10**. Examine the photos with stereolenses and the topographic map to answer the following.

1. What is the contour interval on this map segment? _____

   What is the scale of this map? _____

2. What is the relief between the top of the mesa on the west side of the map (part of Mitchell Mesa) and the

   lower elevations along the east side of the map? _____

3. Of the two Mitten Buttes and Merrick Butte, which is the tallest and what is its elevation?

4. Examining these features in the stereophotos and on the topo map, can you describe any features that indicate differing rock resistances to weathering? (slopes = less resistant, cliffs = more resistance) Describe and

   explain. _____

   _____

   _____

   _____

   _____

   _____

5. The road to the lookout at the campground is paved. Can you find the dirt road on the photo that leaves this

   paved road that goes out across Monument Valley? _____ The campground loop

   road? _____ At what elevation is the visitor center? _____

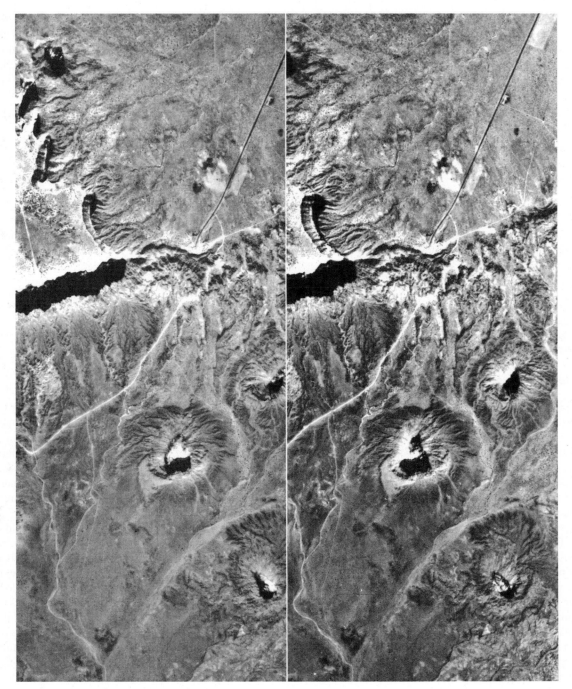

Figure 17.5
Photo stereopair of a portion of Monument Valley, Arizona. In this arrangement, north is to the right, south to the left, and west at the top. (Photos by NAPP, USGS.)

6. Earlier we stated that water is the major erosional force in the desert; although occurring infrequently, when it rains the downpour is usually intense, and floods flash across the sparsely vegetated surfaces swiftly. Given this reality, and using the map and photo stereopairs, describe several examples that demonstrate to you that water is the major erosional force and that it is occurring infrequently. List at least four examples (look for dry washes, sharp features, alluvial deposits).

_____

_____

_____

_____

7. Optional Google Earth™ activity, Mitten Buttes, AZ. For the KMZ file and questions, go to mygeoscience-place.com. Then, click on the cover of *Applied Physical Geography: Geosystems in the Laboratory.*

✳ SECTION 5

Colorado River Water Budget and Cultural Effects

The Colorado River is an exotic stream. Its headwaters lie in the Rocky Mountains and it flows 2,317 km to the arid desert where a trickle of water disappears in the sand, kilometers short of its former mouth in the Gulf of California. Rainfall in the watershed of the Colorado River varies widely, from 102 cm (40 in.) annually in Rocky Mountain National Park in the Rockies, to 20 cm (8 in.) at Grand Junction, Colorado, to a scant 8.9 cm (3.5 in.) per year at Yuma, AZ.

Over the course of its journey the Colorado River discharge experiences evaporation, sedimentation behind dams, and increasing demands of water users! The water budget of the Colorado River has changed over time and it will continue to change due to the highly variable flows of the river and its relation to global climate change. The creation of Hoover Dam in the 1930s and Glen Canyon Dam in 1963 dramatically decreased the transport of sediment and increased the amount of evaporation from expansive reservoir surfaces. The dams created artificial base levels that act to trap sediment that decreases the available storage capacity of the lakes. Glen Canyon Dam formed Lake Powell, which lies on the porous Navajo sandstone. In addition, the lake is an open body of water in an arid desert where hot, dry winds accelerate evaporation. The dams affected the Colorado River by disturbing the pattern of seasonal flows, forming now-permanent sandbars and invasive phreatophytes, which transpire much water from the system.

Six of the seven basin states signed the Colorado River compact in 1923. Arizona and Mexico were added in 1944. Colorado River flows from 1914 to 1923 were used to allocate water rights. Tree ring analysis shows that the only other time Colorado River discharges were as high as the 1914–1923 period was from 1606 to 1625. It was thought that flows were sufficient for the upper and lower basins to receive an absolute amount of 7.5 maf each (million acre feet—the amount of water required to cover one million acres with one foot of water), and for Mexico to receive 0.5 maf. Table 17.1 shows the average flows of the Colorado River at Lees Ferry.

Table 17.1
Average Flows at Lees Ferry

1906–1930	17.70 maf[a]
1930–2003 average flow of the river	14.10 maf
1990–2006	11.34 maf
2000	11.06 maf
2001	10.75 maf
2002	3.1 maf
2003	6.4 maf
2004	5.6 maf
2005	8.3 maf
2006	8.4 maf
2007	8.4 maf
2008	9.2 maf
2009	8.4 maf
2010	8.4 maf
A water year runs from October 1 to September 30.	
[a]1 million acre-feet; 1 acre-foot = 325,872 gallons; 1.24 million liters; 43,560 ft³.	

1. What were the average flows of the Colorado River at Lees Ferry for 1914–1923? What were the average

   flows for 1906–1930? _____ 1990–2006? _____ 2000–2010?

   _____

   What was the maximum flow recorded since 2000? _____ Minimum flow? _____

   _____

   Compare the minimum and maximum flows since 1906 with the water allocation amounts in the treaty and

   compact agreements. _____

   _____

   _____

   _____

2. Why do you suppose water rights were allocated using absolute amounts? What are some of the problems

   with this method of allocation? _____

   _____

   _____

   _____

   _____

3. If you were in charge, how would you allocate water rights to the seven basin states and Mexico? What data would you use? What time periods would you consider, both in the past and in the future?

_____

_____

_____

_____

_____

The Colorado River runs through an area with rapid population growth. California, Arizona, and Nevada all have grown rapidly in the past several decades and their growth rates are remaining high. The population in Nevada is an example, with Las Vegas growing from 368,000 in 1985 to 1.9 million in 2009, an increase of 530%! Population growth alone is putting increasing demands on the Colorado River. The American West has experienced many episodes of drought. Since A.D. 1226 there have been nine droughts lasting more than 20 years each.

The severity of the current western drought that began in 1999 surpasses anything in the historical record and is approaching the driest in the tree-ring record as well. It is estimated that 13 "normal" winters would be required to restore the water levels. Global climate change may increase the severity and frequency of droughts. The subtropical high dominates the weather of the American Southwest and is predicted to intensify and shift northward which would increase the extent of arid conditions, possibly affecting the entire basin.

Table 17.2
Estimated Colorado River Budget through 2010

Water Demand	Quantity (maf)[a]
Upper Basin (7.5) Lower Basin (7.5)[b]	15.0
Central Arizona Project (rising to 2.8 maf)	1.0
Mexican allotment (1944 Treaty)	1.5
Evaporation from reservoirs	1.5
Bank storage at Lake Powell	0.5
Phreatophytic losses (water-demanding plants)	0.5
Budgeted total demand	**20.0 maf**

4. Table 17.2 lists the losses due to evaporation, bank storage (water lost to the porous rock), and phreatophytes (literally water-loving, these are deep-rooted plants with high water needs). The average four-person household uses 1 acre-foot of water per year. The water lost to evaporation, bank storage, and phreatophytes

could serve how many households? _____

_____

5. What percent of the total flow would be lost to evaporation, bank storage, and phreatophytes, given the

1930–2003 average flow? _____ Given the 2002 flow? _____ Given the

2010 average flow? _____

6. What do you think is the future outlook for Colorado River water use (demand) in light of these reductions in discharge (supply)? How would you characterize the pattern of discharge over the past 100 years? Past 20 years? What is the trend in water demand from the river over Arizona, Nevada, and southern California?

   How is global climate change impacting the Colorado River water budget? _____

   _____

   _____

   _____

   _____

7. What strategies would you recommend to help try to balance the water demand for the Colorado River with

   its actual water supply? _____

   _____

   _____

   _____

Name: _____   Laboratory Section:_____

Date: _____   Score/Grade: _____

Lab Exercise 18

Topographic Analysis: Karst Landscapes

Limestone is so abundant on Earth that many landscapes are composed of it. These areas are quite susceptible to chemical weathering. Such weathering creates a specific landscape of pitted and bumpy surfaced topography, poor surface drainage, and well-developed solution channels (dissolved openings and conduits) underground. Remarkable labyrinths of underworld caverns also may develop owing to weathering and erosion caused by groundwater.

These are the hallmarks of **karst topography**, named for the Krš Plateau in Slovenia (formerly Yugoslavia), where karst processes were first studied. Approximately 15% of Earth's land area has some karst features, with outstanding examples found in southern China, Japan, Puerto Rico, Cuba, the Yucatán of Mexico, Kentucky, Indiana, New Mexico, and Florida. As an example, approximately 38% of Kentucky has sinkholes and related karst features noted on topographic maps.

For a limestone landscape to develop into karst topography, there are several necessary conditions:

- The limestone formation must contain 80% or more calcium carbonate for solution processes to proceed effectively.

- Complex patterns of joints in the otherwise impermeable limestone are needed for water to form routes to subsurface drainage channels.

- There must be an aerated (containing air) zone between the ground surface and the water table.

- Vegetation cover supplies varying amounts of organic acids that enhance the solution process.

The role of climate in providing optimum conditions for karst processes remains in debate, although the amount and distribution of rainfall appears important. Karst occurs in arid regions, but it is primarily due to former climatic conditions of greater humidity.

Areas of karst may exhibit diversion of surface water to underground flows. A stream may appear and disappear along its channel. Groundwater and surface waters dynamically interact through the jointed rock structures and permeable rock. Lab Exercise 18 features two sections and two Google Earth™ activities.

Key Terms and Concepts

karst topography
sinkhole

Objectives

After completion of this lab, you should be able to:

1. *Analyze* and *describe* karst landscapes using topographic maps and a photo stereopair.
2. *Relate* these surface features to groundwater and the underground world of solution caverns.

Materials/Sources Needed

pencil
calculator
ruler
stereolenses

Lab Exercise and Activities

SECTION 1

An idealized karst landscape with karst topography

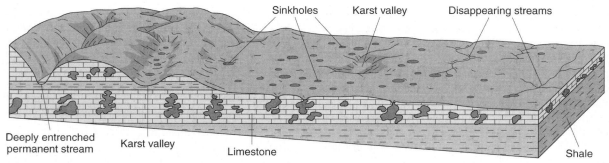

Figure 18.1
Idealized features of a karst landscape

Oolitic, Indiana Quadrangle and Stereophotos

The weathering of limestone landscapes creates many **sinkholes**, which form in circular depressions. (Traditional studies may call a sinkhole a *doline*.) If a solution sinkhole collapses through the roof of an underground cavern, a *collapse sinkhole* is formed. A gently rolling limestone plain might be pockmarked by sinkholes with depths of 2 to 100 m (7 to 330 ft) and diameters of 10 to 1000 m (33 to 3300 ft). Using **Topographic Map #11** and the photo stereopair of part of the mapped area, answer the following.

1. Refer to the Oolitic, Indiana, topographic quadrangle—**Topographic Map #11**. What is the scale of this

   map? _____

   What is the contour interval? _____

2. In the space provided draw the topographic map symbol that indicates a sinkhole; sketch a contour line and illustrate this symbol (check out the chart of topographic map symbols inside the front cover of this manual).

3. Find Section 31 on the map (west of the highway, west center of map). Section 31 is marked on the north and east by a red section line, on the south by a black-dashed township line, and on the west by 500 West Road.

Answer the following about Section 31:

4. What are the highest and lowest elevations in Section 31? Highest: _____

   Lowest: _____

5. How many sinkholes are there in the 1 mi^2 section? _____ . Do any of these sinkholes have water in them (implying that they must have impermeable soil or rock along their base)?

   _____

6. Describe any economic activity specific to this karst environment that you see on the topo map. _____

   _____

   _____

   Examine the photo stereopair with stereolenses on the next page in Figure 18.2 for this same area around Oolitic, Indiana.

7. Describe how the sinkholes appear in the photos. _____

   _____

   _____

8. Can you distinguish between older and newer mining areas by using the stereophotos? How were you able to

   determine this observation? Discuss. _____

   _____

   _____

9. This limestone region is deeply dissected by Salt Creek flowing along the east side of the map and photos. The entrenched meanders east of Oolitic are remarkable in their geometric (rock-structure controlled) bends. Such an entrenched stream is important in the development of a karst region because it permits a continuous movement of water through the jointed limestone landscape. What is the local relief between the town and

   the creek? Examine the contour lines carefully—they are tightly spaced along the cliffs. _____

   _____

   _____

   _____

10. Optional Google Earth™ activity, Oolitic, Indiana. For the KMZ file and questions, go to mygeoscience-place.com. Then, click on the cover of *Applied Physical Geography: Geosystems in the Laboratory*.

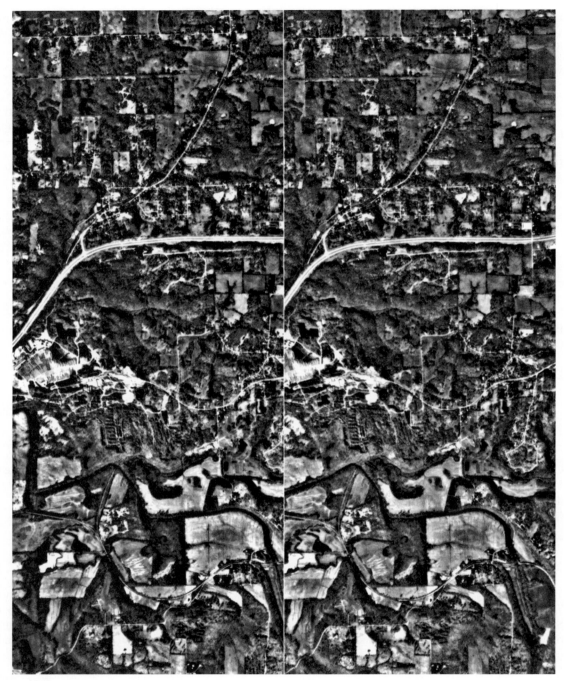

Figure 18.2
Photo stereopair of an area west of Oolitic, Indiana. In this arrangement, north is to the right, west to the top. (Photos by NAPP, USGS.)

Lake Wales, Florida Topographic Quadrangle Map

In Florida, several sinkholes have made news because lowered water tables (lowered by pumping) caused their collapse into underground solution caves, taking with them homes, businesses, and even new cars from an auto dealership. One such sinkhole collapsed in a suburban area in 1981 and others in 1993 and 1998. The area around Lake Wales, Florida, (south of Orlando and north of the Everglades) is characteristic of limestone with karst topography, high water tables, some dry depressions, poor surface drainage, marsh lands, and groves of citrus trees—see **Topographic Map #12**. Complete the following items relative to this map.

1. Refer to the Lake Wales, Florida, topographic quadrangle—**Topographic Map #12**. What is the scale of this

   map? _____

   What is the contour interval? _____

2. How many sinkhole depressions contain lakes in the map segment? _____

3. Do any of these lakes sit in a depression (give a count) 10 ft. deep? _____ 15 ft. deep?

   _____ 20 ft. deep? _____ More than 25 ft.? _____

4. Water will flow downslope between sinkholes. Using lake- and pond-surface elevation as an indicator, in

   which direction do you think groundwater flows through this area? Why? _____

   _____

   _____

5. There is a hospital shown south (SSE) of Lake Wales. Determine the elevation of the hospital as well as

   the elevation of the lake surface of Lake Wales.

   Hospital: _____ Lake Wales: _____

   If you were standing at the hospital, could you see the lake? Explain. _____

   _____

   _____

6. How many dry depressions do you count in this topo map? _____

   These indicate that the bottom of the sinkhole is permeable and cannot hold water permanently, or perhaps the water table lowered possibly due to groundwater mining. When the water table level is above the bottom of the sinkholes, they fill with water.

7. Do you see any evidence that there is a retirement community in Lake Wales (trailer parks or small

   subdivisions of houses in a compact arrangement)? Locate and describe. _____

   _____

   _____

8. Do you find any surface streams on the topographic map? _____. Explain your observation. What

would produce this hydrologic situation? _____

_____

_____

9. Optional Google Earth™ activity, Lake Wales, FL For the KMZ file and questions, go to mygeoscience-place.com. Then, click on the cover of *Applied Physical Geography: Geosystems in the Laboratory.*

Name: _____   Laboratory Section:_____

Date: _____   Score/Grade: _____

Lab Exercise 19

Soils

Earth's landscape generally is covered with soil. **Soil** is a dynamic natural material composed of fine particles in which plants grow, and it contains both mineral fragments and organic matter. The soil system includes human interactions and supports all human, other animal, and plant life. If you have ever planted a garden, tended a house plant, or been concerned about famine and soil loss, this lab exercise will interest you. You may discover that you have made some casual observations that can be applied in this exercise.

Soil science is interdisciplinary, involving physics, chemistry, biology, mineralogy, hydrology, taxonomy, climatology, and cartography. Physical geographers are interested in the spatial patterns formed by soil types and the environmental factors that interact to produce them. **Pedology** concerns the origin, classification, distribution, and description of soil (*ped* from the Greek pedon, meaning "soil" or "earth"). Pedology is at the center of learning about soil as a natural body, but it does not dwell on its practical uses. *Edaphology* (from the Greek *edaphos*, meaning "soil" or "ground") focuses on soil as a medium for sustaining higher plants. Edaphology emphasizes plant growth, fertility, and the differences in productivity among soils. Pedology gives us a general understanding of soils and their classification, whereas edaphology reflects society's concern for food and fiber production and the management of soils to increase fertility and reduce soil losses.

Observing soils first hand in the field is the best way to learn about their properties: construction sites, excavations on your campus or nearby, and roadcuts along highways all provide opportunities for seeing soils in their natural surroundings. In many locales, an *agricultural extension service* can provide specific information and perform a detailed analysis of local soils. Soil surveys and local soil maps are available for most counties in the United States and for the Canadian provinces. Your local phone book may list the U.S. Department of Agriculture, Natural Resources Conservation Service (http://www.nrcs.usda.gov/), or Agriculture Canada's Soil Information System (http://sis.agr.gc.ca/cansis/intro.html, with more information at http://www.metla.fi/info/vlib/soils/old.htm).

This exercise by no means makes you an expert in soil science, but it gives you the opportunity for some hands-on experience with soils, and for using some of the tools and methods that soil scientists use in their work. Lab Exercise 19 features six sections. (*Note: The results for many activities in this exercise will vary, depending on the samples, local sites, and other materials available to your lab.*)

Key Terms and Concepts

clay
humus
loam
pedology
pedon
permeability
polypedon
porosity
sand
silt

soil
soil classification
soil color
soil consistence
soil horizon
soil pH (acidity-alkalinity)
soil profile
soil properties
soil science
soil texture

Objectives

After completion of this lab, you should be able to:

1. *Identify* basic components of soil and soil properties.
2. *Determine* main components of soil sample by color.
3. *Identify* major soil texture categories and *classify* soils by texture.
4. *Use* the "feel method" and *determine* texture of soil samples.
5. *Demonstrate* and *observe* how soil texture affects porosity and movement of water through soil.
6. *Discern* the characteristics and properties of horizons in soil profiles and *identify* horizons.
7. *Measure* pH level in soil samples and *determine* the soil pH (acidity or alkalinity).
8. *Identify* specific soil series that are featured as state soils.

Materials/Sources Needed

pencils
color pencils
ruler
soil samples
Munsell Color Chart (optional, 175 colors)
glass jars or beakers
water
soup can with both ends removed (optional)
pH test equipment (hand held meter or color test kit)
soil samples, or soil sample photographs, or Internet site of soil profiles
Internet access (optional)

Lab Exercise and Activities

❊ SECTION 1

Soil Color

Soil properties are the characteristics or traits of soil, some of which include soil **color, texture**, structure, **consistence, porosity**, moisture, and chemistry. We examine a few of these properties, beginning with color.

Soil color is one of the most obvious traits, suggesting composition and chemical makeup in mineral soils. If you look at exposed soil, color may be the most obvious trait. Among the many possible hues are the reds and yellows found in soils of the southeastern United States (high in iron oxides); the blacks of prairie soils in portions of the U.S. grain-growing regions and Ukraine (richly organic); and white-to-pale hues found in soils containing silicates and aluminum oxides. Reduced iron imparts gray and greenish colors. In dry areas, calcium carbonate or other water-soluble salts will give the soil a white color.

However, color can be deceptive: Soils of high humus content, organic materials from decomposed plant and animal litter, are often dark, yet clays of warm-temperate and tropical regions with less than 3% organic content are some of the world's blackest soils.

To standardize color descriptions, soil scientists describe a soil's color by comparing it with a *Munsell Color Chart* (developed by artist and teacher Albert Munsell in 1913). These charts display 175 colors arranged by *hue* (H, the dominant spectral color, such as red), *value* (V, degree of darkness or lightness), and *chroma* (C, purity and saturation of the color, which increase with decreasing grayness).

Let's examine some essentials relative to soil color in four soil samples.

1. Using soils samples (observed/collected in the field or provided by your instructor, or soil photographs), note the predominant soil color and indicate the likely soil component responsible for the color. (Answers will vary depending on the samples/photographs provided.) Be sure and note whether the sample is wet, moist, or dry.

   a) Soil sample A _____

   _____

   b) Soil sample B _____

   _____

   c) Soil sample C _____

   _____

   d) Soil sample D _____

   _____

Using These Same Four Samples:

In the Munsell color system every color has three qualities: *hue* (the dominant spectral color, such as red), *value* (degree of darkness or lightness), and *chroma* (purity or saturation of the color, which increase with decreasing grayness). The complete Munsell notation for a chromatic color is written symbolically like this: H V/C. As an example, for a strong red having a *hue* of 5R, a *value* of 6, and a *chroma* of 14, the complete Munsell notation is 5R 6/14 (5R is hue, 6 is value, and 14 is chroma). Another example, a pale brown is 10YR 6/3. A dark brown is noted as 10YR 2/2. More refined divisions of any of the attributes, use decimals.

   The light you use when you view the sample is important and can affect your assessment of the color notation. It is best to view the chart and the sample with the Sun over your shoulder shining on the sample, with you facing away from the Sun. If you are in a lab or classroom under artificial light, try to have a light source as close to white light (all spectrum) as possible. Don't be too concerned at this stage, more practice will improve your assessment under varying light conditions. You will find low values and low chromas the most difficult to match against the color chips.

2. Using a *Munsell Color Chart*, note the *color description* and the *Munsell notation* for each of the four samples you gave a preliminary assessment to in the previous response.

   a) Soil sample A _____

   _____

   b) Soil sample B _____

   _____

   c) Soil sample C _____

   _____

   d) Soil sample D _____

   _____

Note: When doing actual field work with a soil pedon (the complete **soil profile** and basic sampling unit in soil surveys), you will find different colors in each horizon, and maybe more than one color in a single horizon. These details should be noted in your assessment.

Soil Texture and Soil Structure

Soil texture refers to the mixture of sizes of its individual particles and the proportion of different sizes of soil separates (individual particles of soil). Figure 19.1 illustrates particle size comparisons and Table 19.1 lists the standards for soil particles grades (sizes). Particles smaller than gravel are considered part of the soil, while larger particles such as gravel, pebbles, or cobbles are not. As you can see from the table, sand is further graded—ranging from very coarse to very fine.

If you have been to a beach, you have felt the texture of **sand**: It has a "gritty" feel. **Silt**, on the other hand, feels smooth—somewhat soft and silky, like flour used in baking bread. When wet, **clay** has a sticky feel and requires quite a bit of pressure to squeeze it, like the clay used in making pottery.

Soils nearly always consist of more than one particle size. By determining the relative amounts of sand, silt, and clay in a particular soil sample, it can be placed into one of twelve classes as shown on the soil texture triangle in Figure 19.2. Each side presents percentages of a particle grade. See the line from each side of the triangle (following the direction indicated by the orientation of the numbers on each axis). You see that a soil consisting of 36% sand, 43% silt, and 21% clay is classified as **loam**, a term for soils consisting of mostly sand and silt with a relatively smaller amount of clay. This determination is a mechanical analysis or particle size analysis of the soil.

Soil structure refers to the arrangement of these soil separates. The smallest natural lump or cluster of particles is a *ped*. Soil structure is described as crumb or granular, platy, blocky, or prismatic or columnar.

Keep this distinction in mind when assessing soil: Soil separates of individual mineral particles are the *soil texture*, and peds, which are the arrangement of soil particles, comprise the *soil structure*.

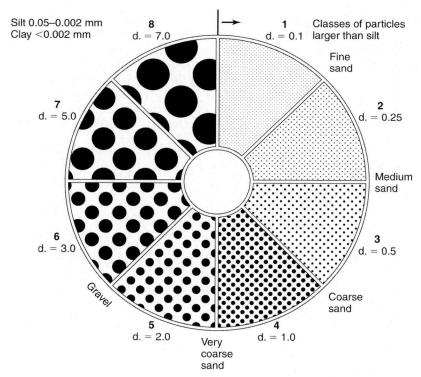

Figure 19.1
Sizes of soil particles larger than silt in approximate diameters (d) of millimeters (mm). (Adapted from *Soil Survey Manual*, USDA, Handbook No. 18, Natural Resources Conservation Service, October 1993, p. 137.)

Table 19.1
Soil Texture Grades
(U.S. Department of Agriculture, Natural Resources Conservation Service)

Soil Particle Grade	Diameter (mm)	Diameter (in)
Gravel	>2.0	>0.08
Very coarse sand	1.0–2.0	0.04–0.08
Coarse sand	0.5–1.0	0.02–0.04
Medium sand	0.25–0.5	0.01–0.02
Fine sand	0.10–0.25	0.004–0.01
Very fine sand	0.05–0.10	0.002–0.004
Silt	0.002–0.05	0.00008–0.002
Clay	<0.002	<0.00008

You can see in the table that soil grades from sand to silt to clay, from most-coarse to most-fine mineral particle sizes. Soils that represent the best particle size mix for plant growth are those that balance the three sizes. This is the *loam* described earlier in this section. Remember, pedology gives us a general understanding of soils and their classification, whereas edaphology reflects society's concern for food and fiber production and the management of soils to increase fertility and reduce soil losses.

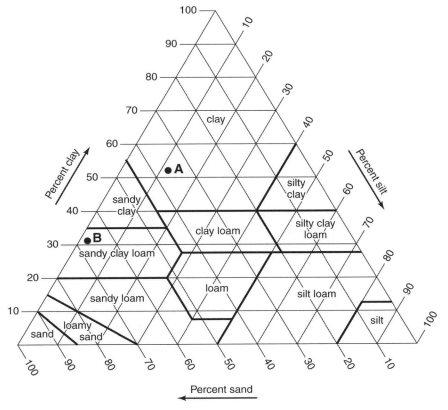

Figure 19.2
Soil texture triangle. (USDA, Natural Resources Conservation Service, same source as Figure 19.1, p. 138.)

1. Use the soil texture triangle to name the following by its correct texture class (named on Figure 19.2):

    a) 17% sand, 28% silt, 55% clay: _____

    b) 31% sand, 55% silt, 14% clay: _____

2. Determine the percentage of each particle size for A and B examples plotted on the triangle (Fig. 18.2):

    a) A: _____ % sand, _____ silt, _____ clay

    b) B: _____ % sand, _____ silt, _____ clay

Please read the appropriate section on soil horizons in *Geosystems* or *Elemental Geosystems*. Soil textures vary through the horizons, so the soil texture triangle can be used for samples in each horizon. For instance, assume a silt loam in Ohio is sampled in the A, B, and C horizons. The mix of mineral particles occurring at each horizon gives you a complete sense of the soil. The textural analysis is as follows:

Table 19.2
Textural analysis of a silt loam soil

Sample points	% Sand	% Silt	% Clay
1 = A horizon	22	63	15
2 = B horizon	31	25	44
3 = C horizon	42	34	24

3. Using the numbers **1, 2,** and **3** to designate each of the three soil horizons, place these numbers in the appropriate location on the soil texture triangle in Figure 19.2. In the textural analysis of mineral particle

    sizes, the A horizon is principally _____; the B horizon is in the majority _____;

    and the C horizon is in the majority _____. In a soil profile is this what you would expect the

    particle size distribution to be? Explain. _____

    _____

    _____

4. In a less quantitative way than using sieves to physically sort soil particles, soil texture can be determined in the field by feeling the soil and estimating the percentages of sand, silt, and clay. Try this method using the following procedure with available soil samples, recording your observations and results through each of these steps. Your instructor may direct you to use the four original soils samples from previous sections of this exercise.

    Step 1: Fill your palm with dry soil, moistening it with enough water so that it sticks together sufficiently to be worked with your fingers. Add the water gradually: if it becomes too runny or if it sticks to your fingers, add more dry soil. You want a "plastic" mass that you can mold, somewhat like putty. (Note: if the soil is gritty and loose, and won't form a ball, but falls apart when rubbed between your fingers, it is a "sandy" soil.)

    Step 2: Knead the soil between your thumb and fingers, removing any pebbles that may be present and breaking up any peds (aggregates or clumps).

    Step 3: The soil should form a ball when you squeeze it. Position the ball between your thumb and forefinger and gently push the ball with your thumb, squeezing it into a ribbon of uniform thickness and width. The ribbon will extend over your finger, eventually breaking from its own weight. Continue until several ribbons have formed and broken off. (If you have difficulty with this, roll the soil into a cylinder that is uniform in diameter, and then squeeze the cylinder of soil into a ribbon.) The *clay content* will determine the length of the ribbon before it breaks:

- ribbon length less than 1.5 in. (3.8 cm): clay content is <27%
- ribbon length 1.5 to 3 in. (3.8 to 7.6 cm): clay content is 27 to 40%
- ribbon length more than 3 in. (7.6 cm): clay content is >40%

Step 4: Now put a small pinch of the soil sample into your palm and add enough water to excessively wet it. Rub it with your forefinger, *estimating* the sand content by the amount of *grittiness* you feel (silt has a smoothness, clay a stickiness in feel):

- sand is the dominant texture you feel: sand content is >50%
- sand is noticeable, but not dominant: sand content is 20–50%
- predominant feel is smooth, indicating high silt content: sand content is <20%

Step 5: Combine the estimates of sand and clay content and use Table 19.3 to place your sample in one of the following classes. (Note: if neither a gritty/sand nor smooth/silt feeling is predominant in Step 4, it will fall into the center column.)

Step 6: Record your observations in the spaces provided after Table 19.3. Your instructor will determine how many samples to use in your work. Space is provided here for three sample reports.

Table 19.3
Simplified Soil Texture Table

		Estimated Percentage of Sand (predominant feeling)		
		>50 (gritty)	20–50 (neither)	<20 (smooth)
Estimated Percentage of Clay	>40	Sandy clay	Clay	Clay
				Silty clay
	27–40	Sandy clay loam	Clay loam	Silty clay loam
	<27	Sandy loam	Loam	Silt loam
		Loamy sand		
		Sand		

Post your results from **Step 4** and **Step 5** and Table 19.3 here:

a) Soil sample A _____

_____

b) Soil sample B _____

_____

c) Soil sample C _____

_____

Note that in soil science, the term *consistence* is used to describe the consistency of a soil or cohesion of its particles. Consistence is a product of texture (particle size) and structure (ped shape). You are working with soil consistence in the exercise above. Consistence reflects a soil's resistance to breaking and manipulation under varying moisture conditions:

- A *wet soil* is sticky between the thumb and forefinger, ranging from a little adherence to either finger, to sticking to both fingers, to stretching when the fingers are moved apart. *Plasticity*, the quality of being moldable, is roughly measured by rolling a piece of soil between your fingers and thumb to see whether it rolls into a thin strand.

- A *moist soil* is filled to about half of field capacity (the usable water capacity of soil), and its consistence grades from loose (noncoherent), to *friable* (easily pulverized), to firm (not crushable between thumb and forefinger).

- A *dry soil* is typically brittle and rigid, with consistence ranging from loose, to soft, to hard, to extremely hard.

✳ SECTION 3

Soil Texture and Porosity

Soil texture affects the movement of water through the soil and, therefore, the amount of water available to plants. Sandy soils have large air spaces between the particles, making them very porous, allowing the water to percolate through the soil quickly. Thus, little water is retained, causing the soil to dry out quickly. You may have observed this at a beach as water quickly drained through the sand, leaving the surface dry and loose.

The small particles in clay soils fit closely together, leaving few air spaces between them so that water percolates very slowly, collecting on the soil surface or running off if there is any slope.

Once the water is absorbed into the soil, it is held so tightly that little is available to the plants.

Therefore, the texture and the structure of the soil dictate available pore spaces, or porosity. The property of the soil that determines the rate of soil-moisture recharge is its permeability. Permeability depends on particle sizes and the shape and packing of soil grains.

Loam soils have a loose, porous texture due to their mixture of particle sizes. As a result, loams retain water well without becoming waterlogged, leaving the water readily available to plants. This is one reason why farmers consider loamy soils ideal. Table 19.4 presents water absorption rates for various soil textures.

Table 19.4
Water Absorption Rates in Soil

Soil Texture	Absorption rate per hour
Sandy	>5 cm (+2 in.)
Loam	0.6–5 cm (0.25–2 in.)
Clay	<0.6 cm (<0.25 in.)

1. You can perform a simple demonstration to observe the effect of soil texture and structure on porosity. Your instructor may have you use the same soil samples from earlier sections in this exercise.

   Step 1: Obtain samples of several soils of differing textures and put each in a glass jar or beaker. Fill the jars to the same level, leaving several inches at the top, taking care not to compact the soil in the container.

   Step 2: Measure 5 cm (2 in.) of water in another container of the same size. Measure the same amount for each sample. Pour the water over the soil in each container and observe how quickly the water percolates through the soil to the bottom of the jar. Roughly time the percolation rates. (Note: the main focus of this demonstration is observation, not exact measurement and quantification of the percolation rates.)

   Step 3: Use the spaces below to make a notation of your observations. Room for three soil sample observations are given. Assess the texture of each sample based on these observations.

a) Soil sample A _____

_____

b) Soil sample B _____

_____

c) Soil sample C _____

_____

2. Optional: A similar demonstration can be done "in the field." Selecting a site where the soil is not com-pacted, trim any vegetation completely to the ground and carefully remove any loose organic materials. Pound a soup can (both ends removed) into the soil several inches. If necessary, place a board over the top of the can to prevent the can from crumpling while you do this. Pour water into the can, filling it to the top. After a period of time, observe how much water has been absorbed. You can do this demonstration at sev-eral sites and compare the results. (Again, while you could take careful measurements, the purpose of this investigation is primarily observation and relative absorption rates.)

a) Site A _____

b) Site B _____

c) Site C _____

❋ SECTION 4

Soil Profiles of Soil Horizons

Just as a book cannot be judged by its cover, so soils cannot be evaluated at the surface only. Instead, a soil profile should be studied from the surface to the deep-est extent of plant roots, or to where regolith or bed-rock is encountered. Such a profile, called a **pedon**, is a hexagonal column measuring 1–10 m^2 in top sur-face area. At the sides of the pedon, the various layers of the soil profile are visible in cross section and are labeled with letters. *A pedon is the basic sampling unit used in soil surveys.*

Many pedons together in one area make up a **polypedon**, which has distinctive characteristics differ-entiating it from surrounding polypedons. A polypedon is comprised of an identifiable series of soils in an area. It can have a minimum dimension of about 1 m^2 and no specified maximum size. *The polypedon is the basic mapping unit used in preparing local soil maps.*

Each distinct layer exposed in a pedon is a **soil horizon**. Each layer or *horizon* is distinct from the one directly above or below, with visible boundaries distinguished by differences in soil properties, some of which are color, texture, porosity, moisture, and the presence or absence of certain moisture, and the pres-ence or absence of certain minerals or other materials.

Figure 19.3 illustrates the horizons in a model or ideal profile. From the surface down the soil horizons are as follows:

• **O horizon**—Named for the organic materials from plant and animal litter. This layer is often subdivided into two layers: the topmost is O_1 consisting of leaves and other largely undecomposed organic material; just be-neath is O_2 which is the decomposed organic debris called **humus**.

• **A horizon**—The uppermost mineral horizon. This dark-colored layer is rich in fine clay-sized particles and organic material derived from the humus above as well as plant roots.

- **E horizon**—Layer named for the process of *eluviation*, in which clays, aluminum and iron oxides, and organic matter are leached (washed out) to lower layers. This horizon is lighter in color, and comprised largely of sand and coarse silt particles.
- **B horizon**—Characterized by the accumulation of materials leached from above. This deposition process is called *illuviation*. Presence of oxides may give this layer reddish or yellowish hues, and the accumulated fine clay and organic particles make this layer quite dense.
- **C horizon**—Not considered part of true soil. Consisting of *regolith* or weathered parent material, this layer lacks biological activity.
- **R horizon**—The bottom layer of the profile, and like the C horizon above, not considered to be part of true soil. This horizon is either unconsolidated (loose) material or solid bedrock.

The photo gallery for many of the orders in the U.S. Soil Taxonomy system, is at **http://soils.usda.gov/gallery/.** The soil profiles pictured at this website might be of use in this section.

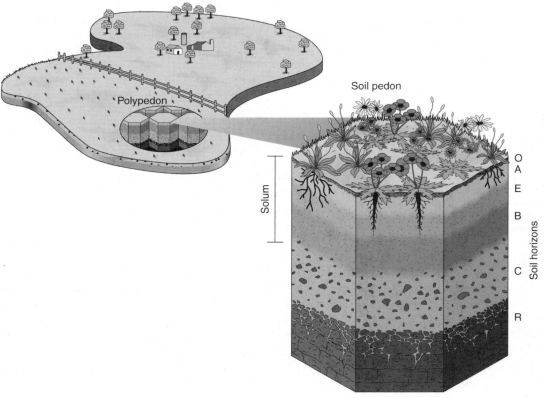

Figure 19.3
Ideal soil profile

1. Using photographs of three soil profiles (either in the soils chapter of a textbook or provided by your instructor) and working in groups with your lab partners, identify the various soil horizons in each profile. Observe horizon characteristics such as thickness, color, texture, and structure, and note variations from one profile to another. Speculate on the factors that combined to produce the appearance of each.

   a) Soil profile A _____

   _____

   b) Soil profile B _____

   _____

c) Soil profile C _____

_____

2. You may be able to locate one or more sites on your campus or nearby where you can see soil profiles in their natural setting. As you did in the first activity, observe the profile's characteristics; if more than one profile is available, note variations between them and see if you can determine the major factors influencing the formation of each.

a) Profile site A _____

_____

b) Profile site B _____

_____

c) Profile site C _____

_____

✳ SECTION 5

Soil Acidity and Alkalinity

Soil fertility is strongly affected by soil acidity or alkalinity as expressed on the pH scale (Figure 19.4). Nutrient availability is low in soils that are either very acidic or very alkaline. A soil solution may contain significant hydrogen ions (H^+), the cations that stimulate acid formation. The result is a soil rich in hydrogen ions, or an acid soil. On the other hand, a soil high in base cations (calcium, magnesium, potassium, sodium) is a basic or alkaline soil.

Pure water is nearly neutral, with a pH of 7.0. Readings below 7.0 represent increasing acidity. Readings above 7.0 indicate increasing alkalinity. Acidity usually is regarded as strong at 5.0 or lower, whereas 10.0 or above is considered strongly alkaline.

Several factors influence soil acidity. The chemistry of soil parent materials, as well as crop fertilization and harvesting can increase soil acidity. However, the major contributor to soil acidity in this modern era is acid precipitation (rain, snow, fog, or dry deposition). Acid rain actually has been measured below pH 2.0—an incredibly low value for natural precipitation, as acid as lemon juice. Increased acidity in the soil solution accelerates the chemical weathering and depletion rates of some mineral nutrients, yet it can also decrease the availability of other nutrients. Because most crops are sensitive to specific pH levels, acid soils below pH 6.0 require treatment to raise the pH. This soil treatment is accomplished by the addition of bases in the form of minerals that are rich in base cations, usually lime (calcium carbonate, $CaCO_3$).

Soil pH is regularly tested at many locations using an *electrometric method* (a probe into a soil-water mixture) or using a *dye method* (organic compounds in drops or test paper that react and indicate pH through color changes). More accurate work can be done in a laboratory. Care must be taken in the field when reading the pH of a soil sample because values may vary over a short distance, perhaps in response to previous fertilizer applications, or reflect topographic effects. Despite these variations, soil pH is related to so many biological and chemical factors that influence both soil processes and plant response, that it is an important parameter to know.

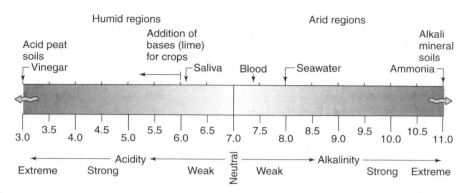

Figure 19.4
pH scale

1. Using pH test equipment provided by your instructor, evaluate two soil samples, perhaps one from your home and one from campus, or from two different areas on your campus. Ideally, you would use samples from sites with different vegetation. Record their pH readings on Figure 19.4 above. How far apart on the

   pH scale are the two readings? _____

2. Are either of the samples strongly acidic? What remedial actions could be taken to make them more pH

   neutral and under what circumstances might you want to do this? _____

3. Review Focus Study 3.2 in *Geosystems* or Focus Study 2.2 in *Elemental Geosystems*. How has acid deposition affected soils, forests, and lakes in the United States and Europe? Do you think that the remedial actions you outlined in item #2 are practical for large-scale problems affecting entire regions? What do you think would be the best solution at these larger spatial scales? (Hint: consider what the sources of acidifying

   materials are.) _____

✳ SECTION 6

[Optional—Internet access required]
In the **soil classification** system developed by the Natural Resources Conservation Service there is a hierarchy of soil categories. Table 19.5 presents the categories of the U.S. Soil Taxonomy.

Table 19.5
U.S. Soil Taxonomy

Soil Category	Number of Soils Included
Orders	12
Suborders	47
Great groups	230
Subgroups	1,200
Families	6,000
Series	15,000

1. Using the map of the world distribution of the Soil Taxonomy systems in *Geosystems*, Figure 18.9, or in *Elemental Geosystems*, Figure 15.8, find your present approximate location. Although this is only a general assessment because of the scale of the world map, use the color map legend and identify the one or two soil orders that seem to represent your region. Record your findings here:

   _____

   _____

2. You can see from Table 19.5 that there are some 15,000 soil series identified in the soil taxonomy for the United States and Canada. In each state a soil series has been selected as representative or as possessing qualities that are of importance in the state. These are displayed on an Internet website by state:

   http://soils.usda.gov/gallery/state_soils/

   To find out about the horizons, color, texture, structure, consistence, mineral and chemical composition of a *state soil* go to this website, prepared by the USDA-Natural Resources Conservation Service.

   Look down the menu and find the state in which you are attending college. The page gives you the soil series name, a representative landscape photo, a soil profile photo of the soil series, a brief description of the soil, a distribution map, and why it was selected to be a state soil. Look for anything special or of note in the soil profile shown. Record the information below. Click on the state name. For instance, go to South Dakota—see something of interest in the B horizon of the soil profile photograph?

   State: _____; **Soil series name:** _____; **Description:** _____

   _____

   _____

   _____

3. Now go to the state of your birth on the menu list of states. If not born in the United States, select any state that is of interest to you. Complete the following, and again be sure to note anything you see of interest in the soil profile photograph shown.

   State: _____; **Soil series name:** _____; **Description:** _____

   _____

   _____

   _____

Note: when doing actual field work with a *soil pedon* (the complete soil profile and the basic sampling unit in soil surveys), you will find different colors in each horizon and perhaps a couple of colors in one horizon. If you identify more than one color in the soil samples provided for this exercise, be sure to note it in your assessment.

Lab Exercise 20

Biomes: Analyzing Global Terrestrial Ecosystems

Many international conferences, treaties, and protocols to protect Earth's air, water, oceans, and climate demonstrate a worldwide concern for the environment and a desire to better understand Earth's ecosystems. From the first Earth Summit in 1992 emerged a new organization—the U.N. Commission on Sustainable Development—to oversee the promises made in the five agreements written at the Earth Summit: the Climate Change Framework; Biological Diversity Treaty; the Management, Conservation, and Sustainable Development of All Types of Forests; the Earth Charter (a nonbinding statement of 27 environmental and economic principles); and Agenda 21 for Sustainable Development. A product of the Earth Summit was the United Nations Framework Convention on Climate Change (FCCC), which led to a series of Conference of the Parties (*COP*) meetings—*COP–13* was in Bali, Indonesia, in 2007, *COP–14* was in Poznan, Poland, in 2008, COP-15 was in Copenhagen, Denmark in 2008, COP-16 was in Cancun, Mexico in 2009, COP-17 will be in Durban, South Africa in late 2011. The Kyoto Protocol, agreed to in 1997 to reduce global carbon emissions, was finalized in 2001, in Marrakech, Morocco, at *COP–7*. Following Russian ratification, the Kyoto Protocol and Rulebook became international law in March 2005, without United States or Australian participation. This momentum also led to Earth Summit 2002 **(http://www.earthsummit2002.org/)** held in Johannesburg, South Africa, with an agenda including climate change, freshwater, gender issues, global public goods, HIV/AIDS, sustainable finance, and the five Rio Conventions.

We need international cooperation to consider our symbiotic relations with each other and with Earth's resilient, yet fragile, life-support systems. A positive step in that direction is the Earth systems science approach embodied in physical geography, that synthesizes content from across the disciplines

to create a holistic perspective. Exciting progress toward an integrated understanding of Earth's physical and biological systems is in progress. Physical geography plays an important role in mapping and analyzing Earth's **terrestrial ecosystems**.

The diversity of organisms is a response to the interaction of the atmosphere, hydrosphere, and lithosphere, producing diverse conditions within which the biosphere exists. A first-ever-international attempt to protect biodiversity (species richness) is now ratified through the auspices of the United Nations.

The biosphere includes myriad ecosystems from simple to complex, each operating within general spatial boundaries. An **ecosystem** is a self-regulating association of living plants and animals and their non-living physical environment. In an ecosystem, a change in one component causes changes in others, as systems adjust to new operating conditions. Interacting populations of plants and animals in an area form a **community.** Each plant and animal occupies an area in which it is biologically suited to live—its habitat—and within that **habitat** it performs a basic operational function—its **niche.**

Earth itself is the largest ecosystem within the natural boundary of the atmosphere. Natural ecosystems are open systems for both energy and matter, with almost all ecosystem boundaries functioning as transition zones rather than as sharp demarcations.

Plants are the most visible part of the biotic landscape, a key aspect of Earth's terrestrial ecosystems. In their growth, form, and distribution, plants reflect Earth's physical systems: its energy patterns; atmospheric composition; temperature and winds; air masses; water quantity, quality, and seasonal timing; soils; regional climates; geomorphic processes; and ecosystem dynamics. The net photosynthesis for an entire plant community (photosynthesis minus respiration) is its *net primary productivity*. This is the amount of stored chemical energy (biomass) that the community generates for the ecosystem. **Biomass** is the net dry weight of organic material and varies among ecosystems.

A large, stable terrestrial ecosystem is known as a **biome.** Specific plant and animal communities and their interrelationship with the physical environment characterize a biome. Each biome is usually named for its *dominant vegetation.* We can generalize Earth's wide-ranging plant species into six broad biomes: *forest, savanna, grassland, shrubland, desert,* and *tundra.* Because plant distributions are responsive to environmental conditions and reflect variation in climatic and other abiotic factors, the world climate map and the global terrestrial biome map are presented together inside the cover of this lab manual.

Earth's diversity is expressed in 270,000 plant species. Despite this complexity of diverse plant and animal communities and their interrelationships, we can generalize Earth's ecosystems into 10 global terrestrial biome regions on the global terrestrial ecosystem map (inside back cover of this lab manual) and in the comprehensive Table 20.1. The table synthesizes many aspects of physical geography, integrating them under the ten biomes. These biomes are included in the glossary, and are presented on the biome map and integrative table. Lab Exercise 20 features four sections and one optional Google Earth™ activity.

Key Terms and Concepts

arctic tundra
biogeography
biomass
biome
boreal forest
community
desert biomes
ecology
ecosystem
equatorial and tropical rain forest
habitat

life zone
Mediterranean shrubland
midlatitude broadleaf and mixed forest
midlatitude grasslands
montane forests
niche
northern needleleaf forest
temperate rain forest
terrestrial ecosystem
tropical savanna
tropical seasonal forest and scrub

Objectives

After completion of this lab, you should be able to:

1. *Define* ecology, biogeography, ecosystem, community, biome, ecotone, and formation class.
2. *Identify* and *differentiate* various terrestrial ecosystem formation classes as they relate to patterns of precipitation and temperature.
3. *Relate* vertical life zones to the latitudinal distribution of biomes.
4. *Compare* Köppen climate classifications from Lab Exercise 11 with the pattern of terrestrial ecosystems (see Appendix B, *Geosystems,* 8/e, or Appendix C in *Elemental Geosystems,* 6/e).
5. *Utilize* a world biome map and an integrative table of major terrestrial biomes and their characteristics to *analyze* environmental conditions.

Materials/Sources Needed

pencil
world atlas or physical geography text

Lab Exercise and Activities

❄ SECTION 1

Essential Terms and Concepts

Define the following and list an example of each:

1. Biogeography: _____

   _____

2. Ecology: _____

   _____

3. Ecosystem: _____

   _____

4. Biome: _____

   _____

5. Community: _____

   _____

6. Habitat: _____

   _____

7. Niche: _____

   _____

❄ SECTION 2

Climate Controls of Ecosystem Structure and Form

Figure 20.1 illustrates the general relationship among temperature, precipitation, and vegetation. **Formation classes** are units that refer to the structure and appearance of dominant plants in a terrestrial ecosystem, for example, **equatorial rain forest**, **northern needleleaf forest**, **Mediterranean shrubland**, and **arctic tundra**. Each formation includes numerous plant communities, and each community includes innumerable plant habitats. The illustration relates temperature and precipitation over hot, temperate, cool, and polar and alpine environments. As you examine the illustration, mentally review what you learned about the characteristics and distribution of climates in Lab Exercise 11.

Questions using Figure 20.1.

1. Describe in general terms the characteristic vegetation type and related temperature and moisture relationship that fit the area of your present town or school. _____

_____

Now, describe the same for the region where you were born (if substantially different). _____

_____

2. Using Figure 20.1, and referring to the world biome map (inside the back cover of this lab manual), write these city names in the appropriate environmental location—locate and label each one on the figure. Discuss these with other lab members to make your determinations.

   a) Your present region

   b) Where you were born (if a different region from your present location)

   c) New York City

   d) Key West, Florida

   e) Montreal, Québec

   f) Dawson, Yukon

   g) Yuma, Arizona

   h) Omaha, Nebraska

   i) Elko, Nevada

   j) Everglades, Florida

   k) Coastal Oregon

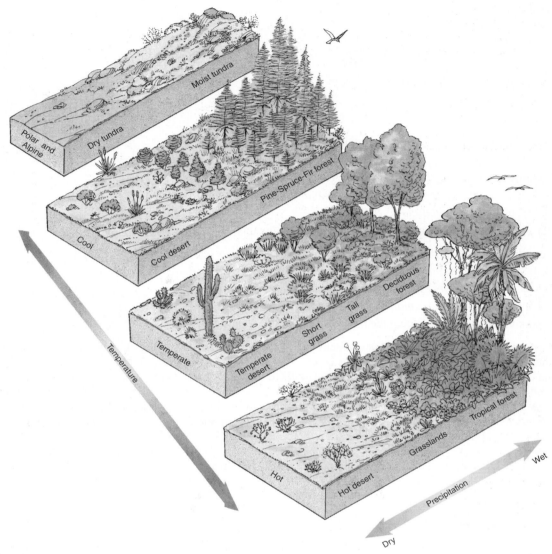

Figure 20.1
Abiotic (nonliving) climate controls of ecosystem types: the generalized relationship among rainfall, temperature, and vegetation.

Recall that Köppen's climatic classification was based on two climate control regimes:

1. **Temperature**—reflecting either
   • Latitude—as in   • Altitude—as in *highland*
   tropical   *or*
   subtropical
   midlatitude
   subarctic/subpolar
   polar
   icecap

2. **Precipitation**—as in
   rainy
   wet and dry
   monsoon
   arid
   semi-arid
   humid
   dry

   In addition, global pressure and wind belts influence precipitation regimes:

- **Rainy/humid conditions**—result from one of the following:
  - instability and convectional uplift in warm *equatorial* and *subtropical* regions
  - frontal uplift in the *midlatitudes*
  - intensification of rainfall on the *windward side* of mountains.

- **Dry conditions**—caused by one of the following:
  - subsiding air of the *subtropical high pressure belts*
  - isolation within *midlatitudes* continental interiors
  - cold stable air of the *high latitudes*
  - drying of air masses descending the *leeward side* of mountain ranges.

These are broad generalizations, but can help you to see the causes of resulting climatic patterns. You will refer to these relationships in ✳ SECTION 4.

The descriptive names in Köppen's climate classification system (see Appendix B, *Geosystems*, or Appendix C in *Elemental Geosystems*) also often included the dominant vegetation types that those regimes supported: *Tropical Rainy or* **Rain forest**, *Tropical Wet and Dry or* **Savanna**, *Tropical/Midlatitude Arid or* **Desert**, *Tropical/Midlatitude Semi-arid or* **Steppe**, *Subtropical Dry or* **Mediterranean**, *Subarctic/Subpolar or* **Taiga**, *Polar or* **Tundra**.

✳ SECTION 3

Life Zones—Conditions Changing with Elevation

Alexander von Humboldt (1769–1859)—an explorer, geographer, and scientist—deduced that plants and animals occur in related groupings wherever similar conditions occur in the abiotic environment. After several years of study in the Andes Mountains of Peru, he described a distinct relationship between elevation and plant communities, his *life zone* concept. As he climbed the mountains, he noticed that the experience was similar to that of traveling away from the equator toward higher latitudes.

This zonation of plants with altitude is noticeable on any trip from lower valleys to higher elevations. Each **life zone** possesses its own temperature, precipitation, and insolation relationships and therefore, its own biotic communities. The key is that temperatures

decrease rapidly with increasing elevation at an average of 6.4 C° per kilometer (3.5 F° per 1000 ft), a rate known as the *normal lapse rate*.

The Grand Canyon in Arizona provides a good example. The inner gorge at the bottom of the canyon (600 m or 2000 ft in elevation) exhibits life forms characteristic of the lower Sonoran Desert of northern Mexico. However, the north rim of the canyon (2100 m or 7000 ft in elevation) is dominated by ecosystems similar to those of southern Canadian forests. On the summits of the nearby San Francisco Mountains (3600 m or 12,000 ft in elevation), the vegetation is similar to the arctic tundra of northern Canada.

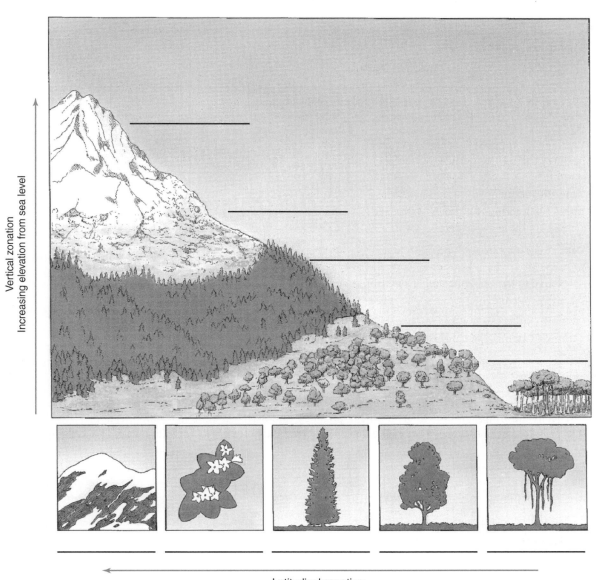

Vertical zonation
Increasing elevation from sea level

Latitudinal zonation
Increasing latitude from the equator

Figure 20.2
Progression of generalized plant community life zones with increasing elevation or latitude.

1. Using a physical geography text, or *Geosystems*, *add* appropriate labels to Figure 20.2 in the spaces provided: tropical rain forest, temperate deciduous forest, needleleaf forest, tundra, and ice and snow. Next, assuming the normal lapse rate is in effect, beginning at sea level and the rain forest label, assume the temperature to be at 27°C (80°F). As you write in each label on the appropriate line provided, assume the labels are spaced at 600 m (2000 ft) intervals, *calculate* the temperature on this particular day at each label elevation. Fill in five labels and calculate five temperatures decreasing with altitude (600, 1200, 1800, 2400, and 3000 m).

2. Following discussion among lab members, compare climate controls (temperature and precipitation) and the effects of altitude—Figures 20.1 and 20.2. Briefly describe the importance of each control on ecosystem character, and plants and animals.

a) temperature: _____

_____

_____

b) precipitation: _____

_____

_____

c) elevation: _____

_____

_____

3. Relative to the life zone concept in Bolivia, South America, speculate on these items:

   a) Given these concepts of elevation and temperature, why are there glaciers along the crest of the Andes

   Mountains so near the equator? _____

   _____

   _____

   b) Why are Bolivians able to grow wheat, potatoes, and barley at 4250 m (14,000 ft) at the same latitude of rain forests at low elevation where these same crops will not grow due to the warmth and dampness?

   _____

   _____

   _____

✳ SECTION 4

Earth's Major Terrestrial Biomes

The global distribution of Earth's major terrestrial biomes is portrayed on the map inside the back cover of this manual. Table 20.1 describes each biome on the map and summarizes other pertinent environmental information—a compilation from many aspects of physical geography, for Earth's biomes are a synthesis of the environment and biosphere.

1. Optional Google Earth™ activity, Deforestation in the Amazon rainforest. For the KMZ file and questions, go to mygeoscienceplace.com. Then, click on the cover of *Applied Physical Geography: Geosystems in the Laboratory.*

Table 20.1
Major terrestrial biomes and their characteristics

Biome and Ecosystems (map symbol)	Vegetation Characteristics	Soil Orders (Soil Taxonomy)	Köppen Climate Designation	Annual Precipitation Range	Temperature Patterns	Water Balance
Equatorial and Tropical Rain Forest (ETR) Evergreen broadleaf forest Selva	Leaf canopy thick and continuous; broadleaf evergreen trees, vines (lianas), epiphytes, tree ferns, palms	Oxisols Ultisols (on well-drained uplands)	Af Am (limited dry season)	180–400 cm (>6 cm/mo)	Always warm (avg 25°C)	Surpluses all year
Tropical Seasonal Forest and Scrub (TrSF) Tropical monsoon forest Tropical deciduous forest Scrub woodland and thorn forest	Transitional between rain forest and grasslands; broad-leaf, some deciduous trees; open parkland to dense undergrowth; acacias and other thorn trees in open growth	Oxisols Ultisols Vertisols (in India) Some alfisols	Am Aw Borders BS	130–200 cm (>40 rainy days during 4 driest months)	Variable, always warm (>18°C)	Seasonal surpluses and deficits
Tropical Savanna (TrS) Tropical grassland Thorn tree scrub Thorn woodland	Transitional between seasonal forests, rain forests, and semiarid tropical steppes and desert; trees with flattened crowns, clumped grasses, and bush thickets; fire association	Alfisols (dry: Ultalfs) Ultisols Oxisols	Aw BS	90–150 cm, seasonal	No cold-weather limitations	Tends toward deficits, therefore fire- and drought-susceptible
Midlatitude Broadleaf and Mixed Forest (MBME) Temperate broadleaf Midlatitude deciduous Temperate needleaf	Mixed broadleaf and needleaf trees; deciduous broadleaf, losing leaves in winter; southern and eastern evergreen pines demonstrate fire association	Ultisols Some Alfisols	Cfa Cwa Dfa	75–150 cm	Temperate, with cold season	Seasonal pattern with summer maximum PRECIP and POTET (PET); no irrigation needed
Needleleaf Forest and Montane Forest (NF/MF) Taiga Boreal forest Other montane forests and highlands	Needleleaf conifers, mostly evergreen pine, spruce, fir; Russian larch, a deciduous needleleaf	Spodosols Histosols Inceptisols Alfisols (boralfs: cold)	Subarctic Dfb Dfc Dfd	30–100 cm	Short summer, cold winter	Low POTET (PET), moderate PRECIP, moist soils, some water-logged and frozen in winter; no deficits
Temperate Rain Forest (TeR) West Coast forest Coast redwoods (U.S.)	Narrow margin of lush ever-green and deciduous trees on windward slopes, red-woods, tallest trees on Earth	Spodosols Inceptisols (mountainous environs)	Cfb Cfc	150–500 cm	Mild summer and mild winter for latitude	Large surpluses and runoff
Mediterranean Shrubland (MSh) Sclerophyllous shrubs Australian eucalyptus forest	Short shrubs, drought adapted, tending to grassy woodlands; chaparral	Alfisols (Xeralfs) Mollisols	Csa Csb	25–65 cm	Hot, dry summers, cool winters	Summer deficits, winter surpluses

(continued)

Table 20.1 (continued)

Midlatitude Grasslands (MGr) Temperate grassland Sclerophyllous shrub	Tallgrass prairies and short-grass steppes, highly modified by human activity; major areas of commercial grain farming; plains, pampas, and veld	Mollisols Aridisols	Cfa Dfa	25–75 cm	Temperate continental regimes	Soil moisture utilization and recharge balanced; irrigation and dry farming in drier areas
Warm Desert and Semidesert (DBW) Subtropical desert and scrubland	Bare ground graduating into xerophytic plants including succulents, cacti, and dry shrubs	Aridisols Entisols (sand dunes)	BWh BWk	<2 cm	Average annual temperature, around 18°C, highest temperatures on Earth	Chronic deficits, irregular precipitation events, PRECIP <½ POTET (PET)
Cold Desert and Semidesert (DBC) Midlatitude desert, scrubland, and steppe	Cold desert vegetation includes short grass and dry shrubs	Aridisols Entisols	BSh BSk	2–25 cm	Average annual temperature around 18°C	PRECIP >½ POTET (PET)
Arctic and Alpine Tundra (AAT)	Treeless; dwarf shrubs, stunted sedges, mosses, lichens, and short grasses; alpine, grass meadows	Gelisols Histosols Entisols (permafrost)	ET Dwd	15–80 cm	Warmest months <10°C, only 2 or 3 months above freezing	Not applicable most of the year, poor drainage in summer
Ice			EF			

Questions and completion items about this sample of Earth's terrestrial biomes.

2. From Table 20.1, the biome map, and the climate map, determine the terrestrial biome that best characterizes each of the following descriptions and write its name *on the first* line provided. *On the second line,* relate it to temperature and precipitation regimes and/or global pressure and wind belts outlined at the end of ❋ SECTION 2. The first one is completed for you as an example.

a) PRECIP less than 1/2 POTET: _____ [*Cold desert and semidesert*] _____

_____ [*midlatitude continental interior and on leeward side of mountain range*] _____

b) Southern and eastern U.S. evergreen pines: _____

_____

c) Mollisols and aridisols: _____

_____

d) Characteristic of central Australia: _____

_____

e) Selva: _____

_____

f) Characteristic of the majority of central Canada: _____

_____

g) Transitional between rain forest and tropical steppes: _____

_____

h) Tallest trees on Earth: _____

_____

i) Sedges, mosses, and lichens: _____

_____

j) Characteristic of Zambia (south central Africa): _____

_____

k) Four biome types that occur in Chile: _____

_____

l) Precipitation of 150–500 cm/year, outside the tropics: _____

_____

m) Characteristic of central Greenland: _____

_____

n) Characteristic of Iran (northeast of the Persian Gulf): _____

_____

o) Major area of commercial grain farming: _____

_____

p) Seasonal precipitation of 90 to 150 cm/year: _____

_____

q) Cfa and Dfa climate types: _____

_____

r) Spodosols and permafrost, short summers: _____

_____

s) Southern Spain, Italy, and Greece, central California: _____

_____

t) Characteristic of Ireland and Wales: _____

_____

u) Just west of the 98th meridian in the United States: _____

_____

v) Just east of the 98th meridian in the United States: _____

_____

w) Bare ground and xerophytic plants: _____

_____

x) East coast of Madagascar: _____

_____

y) West coast of Madagascar: _____

_____

z) Characteristic of northern Mexico: _____

_____

Finally, from the **GEOGRAPHY I. D.** that you completed in the Preface of this lab manual, complete the following for your home town or college campus location:

Place name: _____ Köppen classification: _____

3. What are the main climatic influences for this station (air pressure, air mass sources, degree of continentality,

   temperature of ocean currents)? _____

_____

_____

_____

   Terrestrial biome characteristics: _____

Name: _____

Laboratory Section:_____

Date: _____

Score/Grade: _____

Lab Exercise 21

An Introduction to Geographic Information Systems

Geographic information systems, known by the familiar **GIS** abbreviation are a computer-based, data-processing tools for gathering, manipulating, analyzing, and displaying geographic information. Today's sophisticated computer systems allow the integration of geographic information from direct surveys (on-the ground mapping) and remote sensing in complex ways never before possible. Through a GIS, Earth and human phenomena are analyzed over time. The range of subjects suitable for GIS analysis is limited only by the imagination of the user. If you have used the Internet to get a map or directions, you have used GIS. Today GIS are used to track and predict patterns of disease, help public safety agencies, and plan delivery routes for shipping companies, to name just a few examples. Lab Exercise 21 features three main sections, including one Internet GIS section, and one GIS analysis section.

Key Terms and Concepts

attribute data
buffering
composite overlay
geographic information systems (GIS)
Internet Map Server (IMS)
line
location analysis

point
polygon
raster
spatial data
spatial data structure
vector

Objectives

After completion of this lab, you should be able to:

1. *Describe* essential GIS concepts and *identify* the utility of construction of a GIS model.
2. *Locate* and *utilize* GIS data on the Internet.
3. Use a GIS program to analyze the relationship between earthquakes and plate boundaries.

Materials/Sources Needed

ArcExplorer software
Data for analysis
ESRI's ArcExplorer Java Edition for Education
Internet browser access

Lab Exercise and Activities

❋ SECTION 1

Basic GIS Concepts

Modern GIS owes a great deal to the work of Ian McHarg and his search for an ecologically based land-use planning system. He drew on clear acetate sheets to create layers of information. Each layer showed one theme–roads or land cover or existing houses for example. He stacked multiple sheets to synthesize data and reveal complex relationships. Although he used sheets of plastic to display data (just as you will in the following exercises) rather than a computer screen, the underlying principles are the same. Figure 20.1 is an example of how multiple layers are combined.

GIS combine location information and descriptive information for each feature in a layer. The location information (*spatial data*) tells us where each feature is. The **spatial data** is recorded in a coordinate sys-

tem such as latitude–longitude. The descriptive information (*attribute data*), tells us about the qualities of each feature. **Attribute data** is stored in a spreadsheet or database and is capable of recording multiple attributes for each location.

Spatial data in a GIS are stored as **points** (Figure 21.2a), **arcs** (lines, 21.2b), or **polygons** (areas, 21.2c). Features that only need location information are shown as points. Features that have length are stored as arcs, which are made by combining lines. Finally, features with length and width are stored as polygons. This point-line-polygon system is referred to a **vector data** structure. GIS can also use data stored in another spatial data structure called **raster**. Raster, or grid cells, contain information such as elevation, or remote

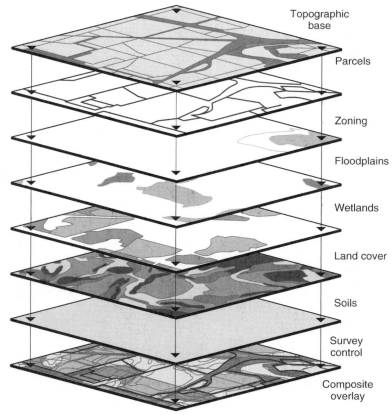

Figure 21.1

Computer-processed spatial data layered in a geographic information system (GIS) produces a composite overlay for analysis. (After USGS.)

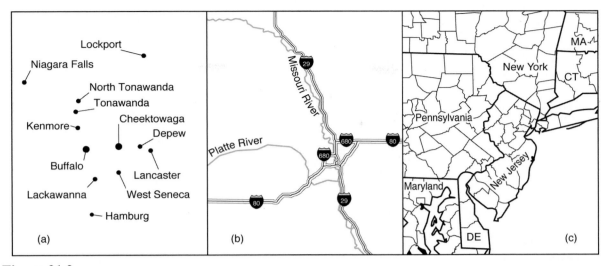

Figure 21.2
Examples of point, line, and polygon features:
cities (a), highways and rivers (b), and states and counties (c)

sensing imagery. Each cell is an intersection of rows and columns and the attribute it contains may occur anywhere in the cell.

By combining layers, a GIS is capable of analyzing patterns and relationships, such as the floodplain or soil layer in Figure 21.1. GIS can create new layers of information by this overlay process as well. A research study may follow specific points or areas through the complex of overlay planes. The utility of a GIS compared with that of a fixed map is the ability to manipulate the variables for analysis and to constantly change the map—the map is alive! Before the advent of computers, an environmental impact analysis required someone to gather data and painstakingly hand-produce overlays of information to determine positive and negative impacts of a project or event. Today, this layered information is handled by a computer driven GIS, which assesses the complex interconnections.

1. Give some examples of GIS that you have used, perhaps without realizing it. _____

_____

_____

_____

_____

2. If you were a land planner, what questions could you answer using the composite overlay of data layers in

Figure 21.1? _____

_____

_____

_____

3. Give examples of features in your life that would be symbolized by points, arcs, and polygons.

Points: _____

Arcs: _____

Polygons: _____

_____

_____

✳ SECTION 2

The Internet and GIS

One of the exciting GIS developments in the last few years is the spread of Internet GIS services. You have probably used many of these services without realizing that you were using a GIS. Some examples would include using the Internet to get directions to a location or to show a map. More sophisticated examples of these are Internet Map Servers (IMS) that allow you to pick which layers are displayed, as well as how they are shown. The main advantage of IMS over conventional GIS is that it does not require special software to be installed on the user's computer, nor does it require special training to use.

An example of a very sophisticated program than can be used with little training is Ushahidi, which uses a web interface to collect and display crowd-sourced data. It was used very successfully after the devastating earthquake in Haiti in 2010. People in Haiti who needed assistance or supplies were able to send a text message to the Ushahidi team. The text messages were received by a team of volunteers all over the world who then entered the data into a GIS that was used to coordinate rescue efforts. The rescue efforts were also assisted by Open-StreetMap (http://www.openstreetmap.org), another crowd-sourced project. OpenStreetMap is a global public street mapping project where volunteers enter and edit street data. The base data can come from remote sensing imagery and users' GPS tracks. During the rescue efforts in Haiti, the command center's location was chosen using OpenStreetMap data. For more information about Ushahidi, visit their website at http://www. ushahidi.com. IMS can also use real-time data, such as traffic speed and accident data to display current road conditions on Google Maps (tm). Some other examples of IMS are:

http://www.arcgis.com/home/gallery.html
An online gallery of user-created maps.
http://www.geomac.gov/
This is an excellent site showing live wildfire information.

The Internet also has opened up access to free data that was either expensive or completely unavailable just a few years ago. For example, California has created the Cal-Atlas Geospatial Clearinghouse at http://atlas.ca.gov, which coordinates GIS data from agencies across the state. This site allows the public to access data easily, and eliminates duplication of effort among agencies. A similar site exists for federal level data at http://www.data. gov/catalog/geodata. Another site with links to data is the Geography Network (http://www.geographynetwork. com/data/downloadable.html).

QUESTIONS:

1. Using the Internet, find three websites that use online GIS. Examples might include real estate, online maps or directions, remote sensing imagery, and wildfire tracking. Record their URLs and descriptions below.

_____

_____

_____

_____

_____

2. Visit three websites that offer free data for downloading that could be used in a GIS. Record their URLs and descriptions below.

_____

_____

_____

_____

✳ SECTION 3

Earthquakes, Volcanoes, and Plate Boundaries

In this section, you will be mapping the locations of earthquakes, volcanoes, and the boundaries of Earth's tectonic plates. You will need to download and install ESRI's ArcExplorer Java Edition for Education, as well as the data for this section. You can find the program at http://www.esri.com/software/arcgis/explorer/arcexplorer.html. The program comes with a tutorial and how-to document. The data and questions for this section can be downloaded from http://www.mygeoscienceplace.com or your instructor may have downloaded it for you already.

Glossary

The Lab Exercise in which each term appears is **boldfaced** and in parentheses, followed by a specific definition related to the usage of the term in *Applied Physical Geography.*

ablation (16) Loss of glacial ice through melting, sublimation, wind removal by deflation, or the calving off of blocks of ice.

actual evapotranspiration (10) Actual amount of evaporation and transpiration that occurs; derived in the water balance by subtracting the deficit from potential evapotranspiration (ACTET).

adiabatic (8) Pertaining to the heating and cooling of a descending or ascending parcel of air through compression and expansion, without any exchange of heat between the parcel and the surrounding environment.

air pressure (7) Pressure produced by the motion, size, and number of gas molecules and exerted on surfaces in contact with the air. Normal sea level pressure, as measured by the height of a column of mercury (Hg), is expressed as 1013.2 millibars, 760 mm of Hg, or 29.92 inches of Hg. Air pressure can be measured with mercury or aneroid barometers.

albedo (8) Reflectivity of a surface. The darker a surface the lower the albedo (the less the surface will reflect the Sun's rays), and the higher the absorption of the Sun's energy.

Alber's projection (14) Alber's equal-area conic projection is used as a base for U.S. quadrangle maps. Two standard parallels are used for the conterminous United States—29.5° N and 45.5° N latitudes. The standard parallels are shifted for conic projections of Alaska (55° N and 65° N) and Hawai'i (8° N and 18° N).

alluvial fan (17) Fan-shaped fluvial landform at the mouth of a canyon, particularly noticeable in arid landscapes where streams are intermittent.

alluvial terraces (17) Level areas that appear as topographic steps above a stream, created by the stream as it scours with renewed downcutting into its floodplain; composed of unconsolidated alluvium (see alluvium).

alluvium (15) General descriptive term for clay, silt, and sand transported by running water and deposited in sorted or semi-sorted sediment on a floodplain, delta, or in a stream bed.

alpine glacier (16) A glacier confined in a mountain valley or walled basin, consisting of three subtypes: valley glacier (within a valley), piedmont glacier (coalesced at the base of a mountain, spreading freely over nearby lowlands), and outlet glacier (flowing outward from a continental glacier).

altimetry (7) The measurement of altitude using air pressure. A *pressure altimeter* is an instrument that measures altitude based on a strict relationship between air pressure and altitude. Air pressure is measured within the altimeter by an aneroid barometer capsule with the instrument graduated in increments of altitude.

altitude (4) The angular distance between the horizon (a horizontal) and the Sun (or any point).

analemma (4) A convenient device to track the passage of the Sun, the Sun's declination, and the positive and negative equation of time throughout the year.

analog (21) Using a different method to portray some type of information. Relating to GIS, the use of manually prepared graphics (e.g., transparencies) to depict spatial information.

aneroid barometer (7) A device to measure air pressure using a partially emptied, sealed cell (see *air pressure*).

anticline (15) The upfolds in a folded landscape, often creating ridges between the downfolded *synclines*.

apparent temperature (5) The temperature subjectively perceived by each individual, also known as sensible temperature.

arctic tundra (20) A biome in the northernmost portions of North America and northern Europe and Russia, featuring low, ground-level herbaceous plants as well as some woody plants.

asthenosphere (12) Region of the upper mantle known as the plastic layer; the least rigid portion of Earth's interior; shatters if struck yet flows under extreme heat and pressure.

attribute table (21) Nonspatial information about a feature being mapped in a GIS (e.g., age, use, type, value), as contrasted to *spatial data.*

available water (10) The portion of capillary water that is accessible to plant roots; usable water held in soil moisture storage.

azimuth (2) A compass direction expressed as an arc area clockwise from the north through 360°.

backswamp (15) A low-lying, swampy area of a floodplain; adjacent to a river, with the river's natural levees on one side and the sides of the valley on the other (see *yazoo tributary*).

barometer (7) A device to measure air pressure; either mercury or aneroid type (see *air pressure*).

barrier spit (17) A depositional form that develops when transported sand in a barrier beach or island is deposited in long ridges that are attached at one end to the mainland and partially cross the mouth of a bay.

bay barrier (17) An extensive sand spit that encloses a bay, cutting it off completely from the ocean and forming a lagoon; produced by littoral drift and wave action; sometimes referred to as a *baymouth bar*.

beach (17) The portion of the coastline where an accumulation of sediment is in motion.

biomass (20) The total mass of living organisms on Earth or per unit area of a landscape; also, the weight of the living organisms in an ecosystem.

biome (20) A large terrestrial ecosystem characterized by specific plant communities and formations; usually named after the predominant vegetation in the region.

boreal forest (20) See *northern needleleaf forests*.

buffering (21) Creating a polygon with a border at a fixed distance around a feature.

capillary water (10) Soil moisture, most of which is accessible to plant roots; held in the soil by surface tension and cohesive forces between water and soil (see also *available water, field capacity*, and *wilting point*).

cardinal points (2) The four principal compass points: north, south, east, and west.

cartography (3) The making of maps and charts; a specialized science and art that blends aspects of geography, engineering, mathematics, graphics, computer science, and artistic specialties.

circle of illumination (4) The division between lightness and darkness on Earth; a day-night great circle.

cirque (16) A scooped-out, amphitheater-shaped basin at the head of an alpine glacier valley; an erosional landform.

classification (11) The process of ordering or grouping data or phenomena in related classes; results in a regular distribution of information; a taxonomy.

climate (11, Prologue) The consistent, long-term behavior of weather over time, including its variability; in contrast to *weather* which is the condition of the atmosphere at any given place and time.

climatic regions (11) Areas of similar climate, which contain characteristic regional weather and air mass patterns.

climatology (11) A scientific study of climate and climatic patterns and the consistent behavior of weather and weather variability and extremes over time in one place or region; and the effects climate and climate change have on human society and culture.

climographs (11) Graphs that plot daily, monthly, or annual temperature and precipitation values for a selected station; may also include additional weather information.

community (20) A convenient biotic subdivision within an ecosystem; formed by interacting populations of animals and plants in an area.

compass points (2) The spots marking the 32 directional points on a compass.

conic (3) A class of map projections based on the geometric shape of a cone.

continental drift (12) A proposal by Wegener in 1912 stating that Earth's landmasses have migrated over the past 200 million years from a supercontinent he called Pangaea to the present configuration; a widely accepted concept today (see *plate tectonics*).

continental glacier (16) A continuous mass of unconfined ice, covering at least 50,000 km^2 (19,500 mi^2); most extensive as ice sheets covering Greenland and Antarctica.

continentality (6) A qualitative designation applied to stations that lack the temperature-moderating effects of the sea and that exhibit a greater range between minimum and maximum temperatures, both daily and annually.

contour interval (14) The vertical distance between adjacent contour lines (see *contour lines*).

contour lines (14) Isolines on a topographic map that connect all points at the same elevation relative to a reference elevation called the vertical datum. The spaces between contour lines portray slope and are known as *contour intervals*.

Coordinated Universal Time (UTC) (1) The official reference time in all countries, formerly known as Greenwich Mean Time outside the United Kingdom; now measured by primary standard atomic clocks whose time calculations are collected in Paris, France, International Bureau of Weights and Measures (BIPM). UTC assumes a prime meridian (0° longitude) passing through Greenwich, London, England.

core (12) The deepest inner portion of Earth, representing one-third of its entire mass; differentiated into two zones—a solid iron inner core surrounded by a dense, molten, fluid metallic-iron outer core.

crust (12) Earth's outer shell of crystalline surface rock, ranging from 5 to 60 km (3 to 38 mi) in thickness from oceanic crust to mountain ranges. Average density of continental crust is 2.7 g per cm^3, whereas oceanic crust is 3.0 g per cm^3.

cylindrical (3) A class of map projections based on the geometric shape of a cylinder.

day length (4) Duration of exposure to insolation, varying during the year depending on latitude; an important aspect of seasonality.

daylight saving time (1) Time is set ahead one hour in the spring and set back one hour in the fall in the Northern Hemisphere. Time is set ahead on the first Sunday in April and set back on the last Sunday in October—only Hawai'i, Arizona, and Saskatchewan exempt themselves. In Europe, the last Sundays in March and September generally are used.

declination (4) The latitude that receives direct overhead (perpendicular) insolation on a particular day; migrates annually through the 47° of latitude between the tropics.

deficit (10) In a water balance, the amount of unmet, or unsatisfied, potential evapotranspiration (DEFIC).

denudation (15) A general term that refers to all processes that cause degradation of the landscape: weathering, mass movement, erosion, and transportation; at the heart of exogenic processes.

desert biomes (20) Arid landscapes of uniquely adapted dry-climate plants and animals.

discharge (13) The measured volume of flow in a river that passes by a given cross-section of a stream in a given unit of time; expressed in cubic meters per second or cubic feet per second.

dew-point temperature (8) The temperature at which a given mass of air becomes saturated, holding all the water it can hold. Any further cooling or addition of water vapor results in active condensation.

drainage basins (15) The basic spatial geomorphic unit of a river system; distinguished from a neighboring basin by ridges and highlands that form divides delimiting the catchment area of the drainage basin, or its watershed.

drainage patterns (15) A geometric arrangement of streams in a region; determined by slope, differing rock resistance to weathering and erosion, climatic and hydrologic variability, and structural controls of the landscape.

drumlin (16) A depositional landform related to continental glaciation; an elongated hill created from till embedded in the bottom of the glacial ice that is plastered, layer upon layer, in a growing mound.

dry adiabatic rate (DAR) (8) The rate at which a parcel of air that is less than saturated cools (if ascending) or heats (if descending); a rate of 10C° per 1000 m (5.5F° per 1000 ft) (see also *adiabatic*).

ecosystem (20) A self-regulating association of living plants, animals, and their nonliving physical and chemical environment.

effusive eruption (12) An eruption characterized by low viscosity, basaltic magma, with characteristic low-gas content readily escaping. Lava pours forth onto the surface with relatively small explosions and little tephra; tends to form shield volcanoes.

empirical classification (11) A classification based on weather statistics or other data; used to determine general climate categories.

environmental lapse rate (8) The actual lapse rate in the lower atmosphere at any particular time under local weather conditions; may deviate above or below the average normal lapse rate of 6.4C° per 1000 m (3.5F° per 1000 ft).

equal area (3) A trait of a map projection; indicates the equivalence of all areas on the surface of the map, although shape is distorted.

equatorial and tropical rain forest (20) A lush biome of tall broadleaf evergreen trees and diverse plants and animals. The dense canopy of leaves is usually arranged in three levels.

equilibrium line (16) The area of a glacier where accumulation (gain) and ablation (loss) are balanced.

equinox (4) March 20–21 and September 22–23 when the circle of illumination divides Earth through the poles and all places have days and nights of equal length.

esker (16) A sinuously curving, narrow deposit of coarse gravel that forms along a meltwater stream channel, developing in a tunnel beneath the glacier. A retreating glacier leaves such ridges behind in a pattern that usually parallels the path of the glacier; may appear branching.

evaporation (10) The movement of free water molecules away from a wet surface into air that is less than saturated; the phase change of water to water vapor; vaporization below the boiling point of water.

evapotranspiration (10) The merging of evaporation and transpiration water loss into one term (see *potential* and *actual evapotranspiration*).

field capacity (10) Water held in the soil by hydrogen bonding against the pull of gravity, remaining after water drains from the larger pore spaces, or storage capacity; the available water for plants. Field capacity is specific to each soil type and is an amount that can be determined by soil surveys.

floodplain (15) A low-lying area near a stream channel, subject to recurrent flooding; alluvial deposits generally mask underlying rock.

fluvial (15) Stream-related processes; from the Latin *fluvius* for "river" or "running water."

formation classes (20) That portion of a biome that concerns the plant communities only, categorized by size, shape, and structure of the dominant vegetation.

gage height (13) The stream depth above stage at maximum peak discharge recorded by a stream gage.

genetic classification (11) A type of classification that uses causative factors to determine climatic regions; for example, an analysis of the effect of interacting air masses on climate.

geographic grid (1) A network of intersecting lines running north–south (through the Poles) and east–west (parallel to the equator) providing a system of global location (see *latitude* and *longitude*).

geographic index number (14) The identification number on a topographic quadrangle, derived from the latitude and longitude of the lower-right corner of the map.

Geographic Information Systems (GIS) (21) A computer-based methodology merging computer cartography with data management, allowing the collection, manipulation, and analysis of geographic/spatial information.

geographic North (or South) Pole (2) The geographic north pole—the point at which all meridians converge—giving a reference point for true north; also known as *true north* (or *true south*).

geography (Preface) The science that studies the interdependence among geographic areas, natural systems, processes, society, and cultural activities over space—a spatial science. The five themes of geographic education include: location, place, movement, regions, and human-Earth relationships.

geomorphology (15) The science that analyzes and describes the origin, evolution, form, classification, and spatial distribution of landforms.

glacier (16) A large mass of perennial ice resting on land or floating shelflike in the sea adjacent to the land; formed from the accumulation and recrystallization of snow, which then flows slowly under the pressure of its own weight and the pull of gravity.

gnomonic projection (3) A planar map projection on which all straight lines are great circle routes.

great circle (1, 3) Any circle of circumference drawn on a globe with its center coinciding with the center of the globe. An infinite number of great circles can be drawn, but only one parallel is a great circle—the equator.

Greenwich Mean Time (GMT) (1) Former world standard time, now known as Coordinated Universal Time (UTC) (see *Coordinated Universal Time*).

habitat (20) That physical location in which an organism is biologically suited to live. Most species have specific habitat parameters, or limits.

hair hygrometer (8) An instrument for the measurement of relative humidity; based on the principle that human hair will change as much as 4% in length between 0 and 100% relative humidity.

heat index (5) The heat index (HI) indicates the human body's reaction to air temperature and water vapor. The level of humidity in the air affects our natural ability to cool through evaporation from skin.

hot spots (12) Individual points of upwelling material originating in the asthenosphere, not necessarily associated with spreading centers, although Iceland is astride a mid-ocean ridge. They tend to remain fixed relative to migrating plates; some 50 to 100 are identified worldwide; exemplified by Yellowstone National Park and Hawai'i (see *plumes*).

humidity (8) Water vapor content of the air. The capacity of the air to hold water vapor is mostly a function of temperature.

humus (19) A mixture of organic debris in the soil, worked by consumers and decomposers in the humification process; characteristically formed from plant and animal litter deposited at the surface.

hurricane (9) A tropical cyclone that is fully organized and intensified in inward-spiraling rainbands; ranges from 160 to 960 km (100 to 600 mi) in diameter, with wind speeds in excess of 119 kmph (65 knots, or 74 mph); a name used specifically in the Atlantic and eastern Pacific (compare *typhoon*).

hydrology (13) The science that studies the flow of water, ice, and water vapor from place to place, as water flows through the atmosphere, across the land, where it is also stored as ice, and within groundwater; including interactions with human activity.

insolation (4) Solar radiation that is intercepted by Earth.

International Date Line (1) The 180° meridian; an important corollary to the prime meridian on the opposite side of the planet; established by the treaty of 1884 to mark the place where each day officially begins.

Internet Map Server (IMS) (21) A combination of hardware and software that allows maps to be displayed and configured over the Internet.

isobars (7) An isoline connecting all points of equal atmospheric pressure.

isogon (2) An isoline connecting points of equal magnetic declination (see *magnetic declination*).

isogonal map (2) A map showing isogons (see *isogons*).

isoline (4) Line on a map connecting all points of equal value.

isotherm (6) An isoline connecting all points of equal air temperature.

kame (16) A small hill of poorly sorted sand and gravel that accumulates in crevasses or in ice-caused indentations in the surface; may also be deposited by water in ice-contact deposits.

karst topography (18) Distinctive topography formed in a region of chemically weathered limestone with poorly developed surface drainage and solution features that appear pitted and bumpy; originally named after the Krš Plateau of Yugoslavia.

kettle (16) Forms when an isolated block of ice persists in a ground moraine, an outwash plain, or valley floor after a glacier retreats; as the block finally melts, it leaves behind a steep-sided hole that may frequently fill with water.

kinetic energy (5) The energy of motion in a body; derived from the vibration of the body's own movement and stated as temperature.

lagoon (17) A portion of coastal seawater that is virtually cut off from the ocean by a bay barrier or barrier beach; also, the water surrounded and enclosed by an atoll.

landfall (9) The location along a coast where a storm moves onshore.

land-water heating differences (6) The differences in the way land and water heat, as a result of contrasts in transmission, evaporation, mixing, and specific heat capacities. Land surfaces heat and cool faster than water and are characterized as having aspects of continentality, whereas water is regarded as producing a moderating marine influence.

latent heat (9) Heat energy is stored in one of three states—ice, water, or water vapor. The energy is absorbed or released in each phase change from one state to another. Heat energy is absorbed as the latent heat of melting, vaporization, or evaporation. Heat energy is released as the latent heat of condensation and freezing (or fusion).

latitude (1) The angular distance measured north or south of the equator from a point at the center of Earth. A line connecting all points of the same latitudinal angle is called a *parallel.*

life zone (20) An altitudinal zonation of plants and animals that form distinctive communities. Each life zone possesses its own temperature and precipitation relationships.

lifting condensation level (8) The height at which an air parcel will reach 100% relative humidity (RH) when it is cooled by dry adiabatic lifting.

line (21) GIS vector feature with one dimension-length.

littoral zone (17) A specific coastal environment that is the region between the high-water line during storms and the depth at which storm waves cannot move sea-floor sediments.

loam (19) A soil that is a mixture of sand, silt, and clay in almost equal proportions, with no one texture dominant; an ideal agricultural soil.

local Sun time (1) Time based on actual longitudinal distance and the prime meridian; also referred to as Sun time or solar time.

longitude (1) The angular distance measured east or west of a prime meridian from a point at the center of Earth. A line connecting all points of the same longitude is called a meridian.

magnetic declination (2) The horizontal angle between geographic (true) north and magnetic north.

magnetic North (or South) Pole (2) The point on Earth's surface indicated by the north-seeking point of a magnetic compass needle.

mantle (12) An area within the planet representing about 80% of Earth's total volume, with densities increasing with depth and averaging 4.5 g per cm^3; occurs above the core and below the crust; is rich in iron and magnesium oxides and silicates.

map, map projection (3) The reduction of a spherical globe onto a flat surface in some orderly and systematic realignment of the latitude and longitude grid.

march of the seasons (5) The annual passage of Earth as it revolves around the Sun producing seasonal changes in daylength and Sun altitude.

marine (6) Descriptive of stations that are dominated by the moderating effects of the ocean and that exhibit a smaller minimum and maximum temperature range than continental stations (see *land-water heating differences*).

Mediterranean shrubland (20) A major biome dominated by shrub formations; occurs in Mediterranean dry summer climates and is characterized by sclerophyllous scrub and short, stunted, tough forests.

Mercator projection (3) A true-shape cylindrical projection on which straight lines are true compass directions (lines of constant bearing)—rhumb lines.

mercury barometer (7) A device that measures air pressure with a column of mercury in a tube that is inserted in a vessel of mercury. The surrounding air exerts pressure on the mercury in the vessel (see *air pressure*).

meridian (1) See *longitude.*

mesosphere (6) The upper region of the homosphere from 50 to 80 km (30 to 50 mi) above the ground; designated by temperature criteria and very low pressures, ranging from 0.1 to 0.001 mb. The top of the mesosphere, as determined by temperature (−90°C) is the **mesopause**.

meteorology (9, Prologue) The scientific study of the atmosphere that includes a study of the atmosphere's physical characteristics and motions, related chemical, physical, and geological processes, the complex linkages of atmospheric systems, and weather forecasting.

mid-ocean ridges (12) Submarine mountain ranges that extend more than 65,000 km (40,000 mi) worldwide and average more than 1000 km (620 mi) in width; centered along sea-floor spreading centers as the direct result of upwelling areas of heat in the upper mantle.

midlatitude broadleaf and mixed forest (20) A biome in moist continental climates in areas of warm-to-hot summers and cool-to-cold winters; includes several distinct communities. Relatively lush stands of broadleaf forests trend northward into needleleaf evergreen stands.

midlatitude grasslands (20) The major biome most modified by human activity; so named because of the predominance of grass-like plants, although deciduous broadleafs appear along streams and other limited sites; location of the world's bread-baskets of grain and livestock production.

moist adiabatic rate (MAR) (8) The rate at which a parcel of saturated air cools in ascent; a rate of 6C° per 1000 m (3.3F° per 1000 ft). This rate may vary, with moisture content and temperature, from 4C° to 10C° per 1000 m (2F° to 6F° per 1000 ft). Also referred to as wet adiabatic rate (WAR) (see *adiabatic*).

montane forests (20) Needleleaf forests associated with mountain elevations (see *northern needleleaf forest*).

moraine (16) Marginal glacial deposits of unsorted and unstratified material, producing a variety of depositional landforms.

natural levees (15) Long, low ridges that occur on either side of a river in a developed floodplain; sedimentary (coarse gravels and sand) depositional by-products of river-flooding episodes.

niche (20) The basic function, or occupation, of a life form within a given community; the way an organism obtains its food, air, and water.

normal lapse rate (5, 8) The average rate of temperature decrease with increasing altitude in the lower atmosphere; an average value of 6.4C° per km, or 1000 m (3.5F° per 1000 ft); a rate that exists principally during daytime conditions.

northern needleleaf forest (20) Forests of pine, spruce, fir, and larch, stretching from the east coast of Canada westward to Alaska and continuing from Siberia westward across the entire extent of Russia to the European Plain; called the taiga (a Russian word) or the boreal forest; principally in the D climates. Includes montane forests.

orographic lifting (8) The uplift of migrating air masses in response to the physical presence of a mountain, a topographic barrier. The lifted air cools adiabatically as it moves upslope; may form clouds and produce increased precipitation (see *rain shadow*).

outwash plain (16) Glacio-fluvial deposits of stratified drift of meltwater-fed, braided, and over-loaded streams with debris beyond a glacier's morainal deposits.

oxbow lake (15) A lake that was formerly part of the channel of a meandering stream; isolated when a stream eroded its outer bank forming a cutoff through the neck of a looping meander.

Pangaea (12) The supercontinent formed by the collision of all continental masses approximately 225 million years ago; named by Wegener in 1912 in his continental drift theory.

parallel (1) See *latitude*.

paternoster lake (16) One of a series of small, circular, stair-stepped lakes formed in individual rock basins aligned down the course of a glaciated valley; named because they look like a string of rosary (religious) beads.

pedon (19) A soil profile extending from the surface to the lowest extent of plant roots or to the depth where regolith or bedrock is encountered; imagined as a hexagonal column; the basic soil sampling unit.

permeability (19) The ability of water to flow through soil or rock; a function of the texture and structure of the medium.

physical geography (Preface) A science that studies the spatial aspects of the physical elements and processes that make up the environment: energy, air, water, weather, climate, landforms, soils, animals, plants, and Earth.

planar (3) A class of map projections based on the geometric shape of a plane. Also known as azimuthal.

planimetric map (14) A basic map showing the horizontal position of boundaries, land-use activities, and political, economic, and social outlines.

plate tectonics (12) The conceptual model that encompasses continental drift, sea-floor spreading, and related aspects of crustal movement; widely accepted as the foundation of crustal tectonic processes.

plumes (12) Columns of heated rock rising through the asthenosphere or mantle, creating hot spots in the migrating over-lying lithospheric plates (see *hot spots*).

point (21) GIS vector feature that consists of a location.

polygon (21) GIS vector feature of an area.

polypedon (19) The identifiable soil in an area, with distinctive characteristics differentiating it from surrounding polypedons that form the basic mapping unit; composed of many pedons (see *pedon*).

porosity (19) The total volume of available pore space in soil; a result of the texture and structure of the soil.

potential evapotranspiration (10) The amount of moisture that would evaporate and transpire if adequate moisture were available; the amount lost under optimum moisture conditions (m) that is, the moisture demand (POTET).

precipitation (10) Rain, snow, sleet, and hail—the moisture supply (PRECIP).

pressure gradient (7) The change in atmospheric pressure horizontally across Earth's surface; measured along a line at right angles to the isobars.

prime meridian (1) An arbitrary meridian designated as 0° longitude; the point from which longitude is measured east or west; agreed on by the nations of the world in an 1884 treaty.

profile view (14) See *topographic profile*.

quadrant compass bearing (2) A compass direction expressed as an arc clockwise or counterclockwise (toward the east or toward the west—whichever is closer) through 90° from the north or the south.

rain shadow (8) The area on the leeward slopes of a mountain range; in the shadow of the mountains, where precipitation receipt is greatly reduced compared to windward slopes (see *orographic lifting*).

raster (21) In GIS, a data structure based on a grid of (usually) square and rectangular "cells."

rating curve (13) Relates stream discharge to gage height in showing the direct relationship between stream discharge and its width, depth (gage height), and velocity.

recurrence intervals (13) The average time period within which a given event will be equaled or exceeded once, also called *return period*.

regime (13) Characteristic seasonal variation in streamflow, reflecting climatic conditions—a stream's annual "personality."

relative humidity (8) A term that reflects the ratio of water vapor actually in the air (content) compared to the maximum water vapor the air is able to hold (capacity) at that temperature; expressed as a percentage.

relief (14) Elevation differences in a local landscape as an expression of the unevenness, height, and slope variation.

revolution (4) The annual orbital movement of Earth about the Sun; determines year length and the length of seasons.

rhumb line (3) A line of constant compass direction, or constant bearing, which crosses all meridians at the same angle. A portion of a great circle.

rotation (4) The turning of Earth on its axis; averages 24 hours in duration; determines day-night relationships.

saturated (8) Air that is holding all the water vapor that it can hold at a given temperature.

scale (3) The ratio of the distance on a map to that in the real world; expressed as a representative fraction, graphic scale, or written scale.

sea-floor spreading (12) As proposed by Hess and Dietz, the mechanism driving the movement of the continents; associated with upwelling flows of magma along the worldwide system of mid-ocean ridges.

seismic waves (12) The shock waves sent through the planet by an earthquake or underground nuclear test. Transmission varies according to temperature and the density of various layers within the planet.

sensible heat (5) Heat that can be measured with a thermometer; a measure of the concentration of kinetic energy from molecular motion (see *apparent temperature*).

sinkhole (18) Nearly circular depression created by the weathering of karst landscapes; also known as a doline in traditional studies; may collapse through the roof of an underground space (see *karst topography*).

sling psychrometer (8) A device for the measurement of relative humidity using two thermometers—a dry bulb and a wet bulb—mounted side-by-side.

small circle (1) Circles on a globe's surface that do not share Earth's center; for example, all parallels other than the equator.

soil (19) A dynamic natural body made up of fine materials covering Earth's surface in which plants grow, composed of both mineral and organic matter.

soil horizon (19) The various layers exposed in a pedon; roughly parallel to the surface and identified as O, A, E, B, C, and R (bedrock).

soil pH (acidity-alkalinity) (19) The relative abundance of free hydrogen ions (H^+) in a solution. The pH scale is logarithmic: Each whole number represents a 10-fold change. A pH of 7.0 is neutral, and values less than 7.0 are increasingly acidic, while values greater than 7.0 are increasingly basic or alkaline.

soil moisture storage; recharge; utilization (10) The retention of moisture within soil; represents a savings account that can accept deposits (*soil moisture recharge*) or experiences withdrawals (*soil moisture utilization*) as conditions change.

soil profile (19) See *pedon*.

soil science (19) An interdisciplinary science, involving physics, chemistry, biology, mineralogy, hydrology, taxonomy, climatology, and cartography, that studies soils.

soil-water budget (10) An accounting system for soil moisture using inputs of precipitation and outputs of evapotranspiration and gravitational water.

solstice (5) June 20–21 and December 21–22 when the Sun's declination is at the tropics of Cancer or Capricorn, respectively.

spatial data (21) In a GIS, the data that is concerned with the location of a feature, as contrasted with *attribute data*.

spatial data structure (21) The way locational/spatial data is organized in a GIS—either raster- or vector-based.

specific heat (6) The increase of temperature in a material when energy is absorbed. Because water requires far more heat to raise its temperature than does a comparable volume of land, water is said to have a higher specific heat.

specific humidity (8) The mass of water vapor (in grams) per unit mass of air (in kilograms) at any specified temperature. The maximum mass of water vapor that a kilogram of air can hold at any specified temperature is termed its maximum specific humidity.

stability (8) The condition of a parcel of air relative to whether it remains where it is or changes its initial position. The parcel is stable if it resists displacement upwards, unstable if it continues to rise; relates to adiabatic and lapse rate processes.

standard atmosphere (7) An agreed-upon model of Earth's atmosphere, with set values for the vertical distribution of temperature, pressure, and density from the surface to approximately 80 km (50 mi).

standard line (3) The line of tangency on a map projection along which the projection surface (cylinder or cone) contacts the globe; along this line the map maintains accuracy, free of distortion.

stereolenses (stereoscope) (Prologue) A device that forces each eye to look straight ahead, thus reproducing binocular vision using a pair of overlapping photographs; permits 3-D viewing of photo stereopairs.

stereoscopic contour map (14) A three-dimensional graphic representation of elevation, viewed through a stereoscope; also known as a stereogram.

storm surge (9) Large quantities of seawater pushed inland by the strong winds associated with a tropical cyclone.

stratosphere (6) That portion of the homosphere that ranges from 20 to 50 km (12.5 to 30 mi) above Earth's surface, with temperatures ranging from −57°C(−70°F) at the tropopause to 0°C(+32°F) at the **stratopause**. The functional ozonosphere is within the stratosphere.

subduction zone (12) An area where two crustal plates collide and the denser oceanic crust dives beneath the less dense continental plate, forming deep oceanic trenches and seismically active regions.

subsolar point (4) The only point receiving perpendicular insolation at a given moment—the Sun directly overhead.

surplus (10) The amount of moisture that exceeds potential evapotranspiration; moisture oversupply when soil moisture storage is at field capacity.

syncline (15) In a landscape formed by folding, the downfold or trough often creating valleys between the upfolded *anticlines*.

synoptic map (9) A weather map that shows a moment in time over a large area.

tarn (16) A small mountain lake which occupies a basin in a glacial trough or cirque.

temperate rain forest (20) A major biome of lush forests at middle and high latitudes; occurs along narrow margins of the Pacific Northwest in North America, among other locations; a mixture of broadleaf and needleleaf trees and thick undergrowth, including the tallest trees in the world.

temperature (5) A measure of sensible heat energy present in the atmosphere and other media, indicates the average kinetic energy of individual molecules within the atmosphere.

terrestrial ecosystem (20) A self-regulating association characterized by specific plant formations; usually named for the predominant vegetation and known as a biome when large and stable.

thermosphere (6) A region of the heterosphere extending from 80 to 480 km (50 to 300 mi), the **thermopause**, in altitude; contains the functional ionosphere layer.

till plains (16) Large, relatively flat plains composed of unsorted glacial deposits behind a terminal or end moraine. Low-rolling relief and interrupted, unclear drainage patterns are characteristic.

tombolo (17) A landform created when coastal sand deposits connect the shoreline with an offshore island outcrop or sea stack.

topographic map (14) A map that portrays physical relief through the use of elevation contour lines that connect all points at the same elevation above or below a vertical datum, such as mean sea level.

topographic profile (14) A graphic representation of graduated elevations along a line segment drawn on a map—a *profile view*—is an important method of analysis for a topographic map.

topography (14) The undulations and configurations that give Earth's surface its texture; the heights and depths of local relief including both natural and human-made features.

township and range (3) The Public Lands Survey System (1785) delineated a plan that called for the establishment of a grid system based on an initial survey point; most of the United States is subdivided into 36 mi^2 townships, 1 mi^2 sections, and 160 acre homesteads (quarter section).

transform faults (12) An elongate zone along which faulting occurs between mid-ocean ridges; produces a relative horizontal motion with no new crust formed or destroyed; strike-slip motion, either left or right lateral.

transpiration (10) The movement of water vapor out through the pores in leaves drawn by their roots from the soil moisture storage.

tropical cyclone (9) A cyclonic circulation originating in the tropics, with winds between 30 and 64 knots (39 to 73 mph); characterized by closed isobars, circular organization, and heavy rains (see *hurricane* and *typhoon*).

tropical savanna (20) A major biome containing large expanses of grassland interrupted by trees and shrubs; a transitional area between the humid rain forests and tropical seasonal forests and the drier, semiarid tropical steppes and deserts.

tropical seasonal forest and scrub (20) A variable biome on the margins of the rain forests, occupying regions of lesser and more erratic rainfall; the site of transitional communities between the rain forests and tropical grasslands.

troposphere (6) The home of the biosphere; the lowest layer of the homosphere, containing approximately 90% of the total mass of the atmosphere; extends up to the **tropopause**, marked by a temperature of $-57°C$ ($-70°F$); occurring at an altitude of 18 km (11 mi) at the equator, 13 km (8 mi) in the middle latitudes, and at lower altitudes near the poles.

true north (2) See *geographic north*.

true shape (3) A map property showing the correct configuration of coastlines; a useful trait of conformality for navigational and aeronautical maps, although areal relationships are distorted.

typhoon (9) A tropical cyclone with inward-spiraling winds in excess of 65 knots (74 mph) that occurs in the western Pacific; same as a hurricane except for location.

Universal Transverse Mercator grid (UTM) (2) A special cylindrical projection in which the projection surface is tangent to the globe at a selected pair of opposing meridians or intersects at two selected small circles; scale is constant along those vertical lines. The UTM grid has become the standard for military applications.

vector (21) A GIS data format where point, line, or polygon files are composed of coordinate pairs of x, y locations.

wind chill factor (5) The wind-chill factor indicates the enhanced rate at which body heat is lost to the air. As wind speeds increase, heat loss from the skin increases.

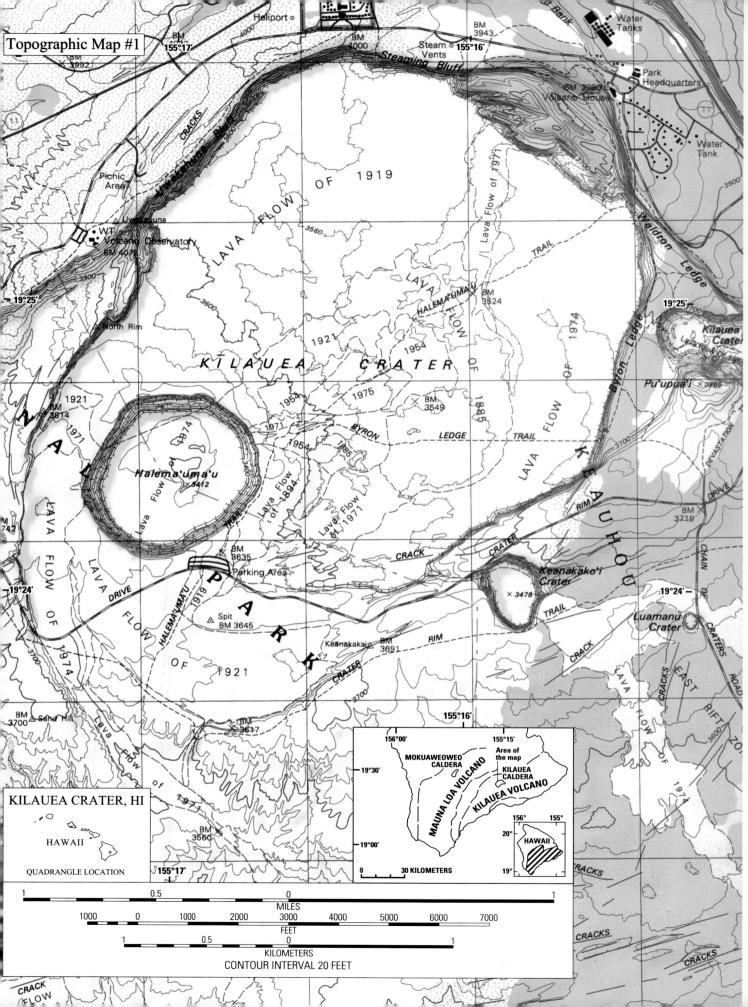

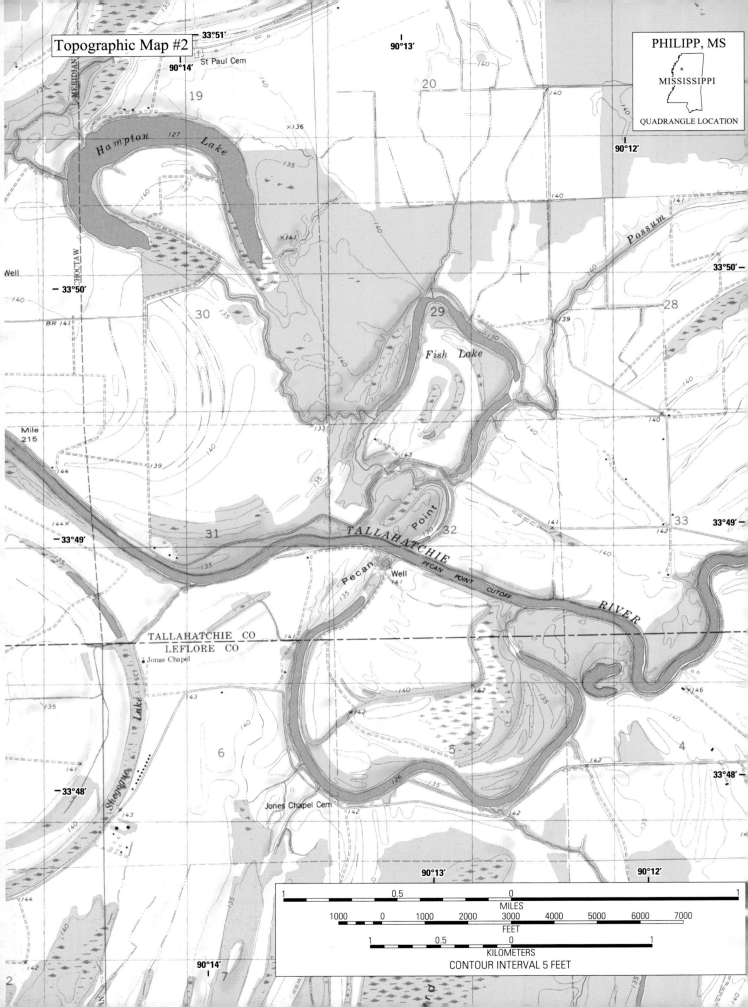

Topographic Map #2

PHILIPP, MS

MISSISSIPPI

QUADRANGLE LOCATION

33°51'
90°14'
St Paul Cem

90°13'

90°12'

90°14'

33°50'

33°49'

33°48'

90°13'

90°12'

MERIDIAN

CHOCTAW

Hampton   Lake

Possum

Fish Lake

TALLAHATCHIE

Point

Pecan   PECAN   POINT   CUTOFF

Well

RIVER

TALLAHATCHIE CO
LEFLORE CO

Jonas Chapel

Sheppard   Lake

Jones Chapel Cem

Well

Mile
215

BR 141

19

20

30

29

28

31

32

33

6

5

4

2

MILES

FEET

KILOMETERS

CONTOUR INTERVAL 5 FEET

1   0.5   0   1

1000   0   1000   2000   3000   4000   5000   6000   7000

1   0.5   1

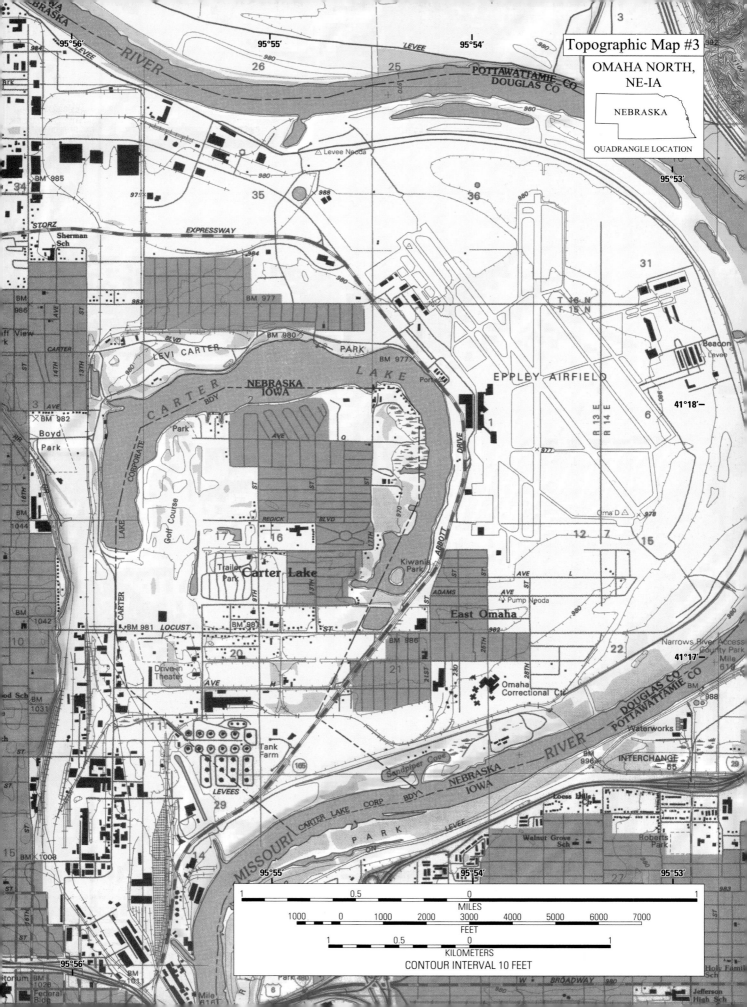

Topographic Map #3

OMAHA NORTH,
NE-IA

NEBRASKA

QUADRANGLE LOCATION

CONTOUR INTERVAL 10 FEET

Topographic Map #4

MILES

FEET

KILOMETERS

CONTOUR INTERVAL 20 FEET

CUMBERLAND, MD

MARYLAND

QUADRANGLE LOCATION

MILES

FEET

KILOMETERS

CONTOUR INTERVAL 40 FEET

Topographic Map #6

121°48' 121°46' 121°44'

MT. RAINIER, WA

WASHINGTON

QUADRANGLE LOCATION

121°42'

46°52'

46°52'

46°50'

46°50'

46°48'

46°48'

46°46'

46°46'

121°48' 121°46' 121°44' 121°42'

MILES
1   0.5   0   1

FEET
1000 0 1000 2000 3000 4000 5000 6000 7000 8000 9000 10000

KILOMETERS
1   0.5   0   1

CONTOUR INTERVAL 80 FEET

Topographic Map #7

JACKSON, MI

MICHIGAN

QUADRANGLE LOCATION

CONTOUR INTERVAL 10 FEET

Topographic Map #8

122°57'30"
123°55'00"
122°52'30"

38°07'30"
38°07'30"

38°05'00"
38°05'00"

38°02'30"
38°02'30"

38°00'00"
38°00'00"

122°57'30"
122°55'00"
122°52'30"

PACIFIC

SAND

BEACH

HIGHWAY

Abbotts Lagoon

McClures Ranch
WT

BM 89
BM 51
Call Building
Quarry
H Ranch

PUNTA

POINT

REYES

DRAKE

Creamery Bay
Bull Point

Schooner Ldg (Site)

Radio Sta
Radio Sta
BM 108
BM 34
Blacksmith Shop (Site)
BM 39
Mud

Schooner Bay

Home Bay

Schooner Ldg (Site)
Mud

DRAKES ESTERO

Mud
Mud
Mud
Mud

Barries Bay
Spring
BM 157
BM 237
BM 323
BM 166
D Ranch
BM 253
BM 116
BM 283
BM
552
BM 217
BM 13

Radio Relay Sta

Chimney Rock

REYES

Drakes Beach

DRAKES

M Ranch
Water Tank
BM 141
BM 59

Water Tank

Home Ranch
Home Ranch

Water Tanks
Water Tank

Water Tank

DE

REYES

LOS

SEASHORE

NATIONAL

Mt Vision
Pt Reyes Hill
Radio Range Sta

Glenbrook

Muddy Hollow

Laguna Ranch

Limantour Spit
Sand
Sand

Coast Campgrd

Reef Sta

Home Ranch Creek
Creek

Limantour
Drakes Head
de Estero
Sand

BM 303
BM 472
BM 334
BM 111
BM 8

STATE PARK

TOMALES BAY

Duck Cove
Marconi
BM 12

Indian Beach
Hearts Desire
Rock
Pebble Beach
Shallow Beach
Shell Beach
Teachers Beach
Oyster Pen
Millert Point

Inverness

PUNTA DE LOS REYES

SAN ANDRE

POINT REYES, CA

CALIFORNIA

QUADRANGLE LOCATION

MILES
1   0.5   0   1   2   3   4

FEET
3000   0   3000   6000   9000   12000   15000   18000   21000

KILOMETERS
1   0.5   0   1   2   3   4   5

CONTOUR INTERVAL 80 FEET

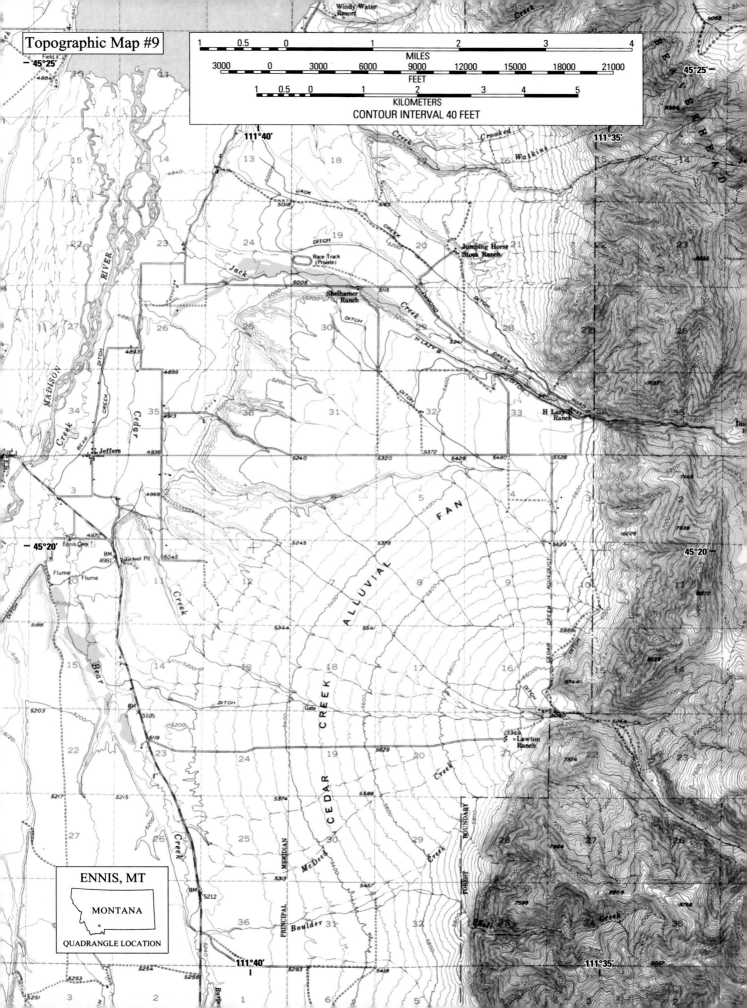

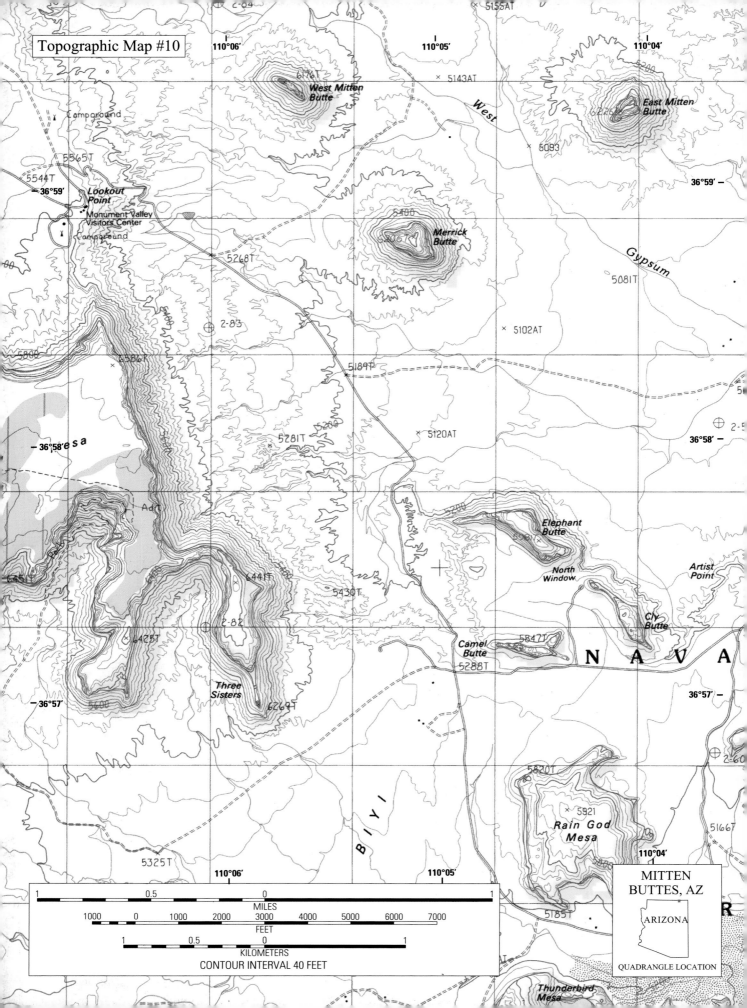

Topographic Map #11

OOLITIC, IN

INDIANA

QUADRANGLE LOCATION

MILES

FEET

KILOMETERS

CONTOUR INTERVAL 10 FEET

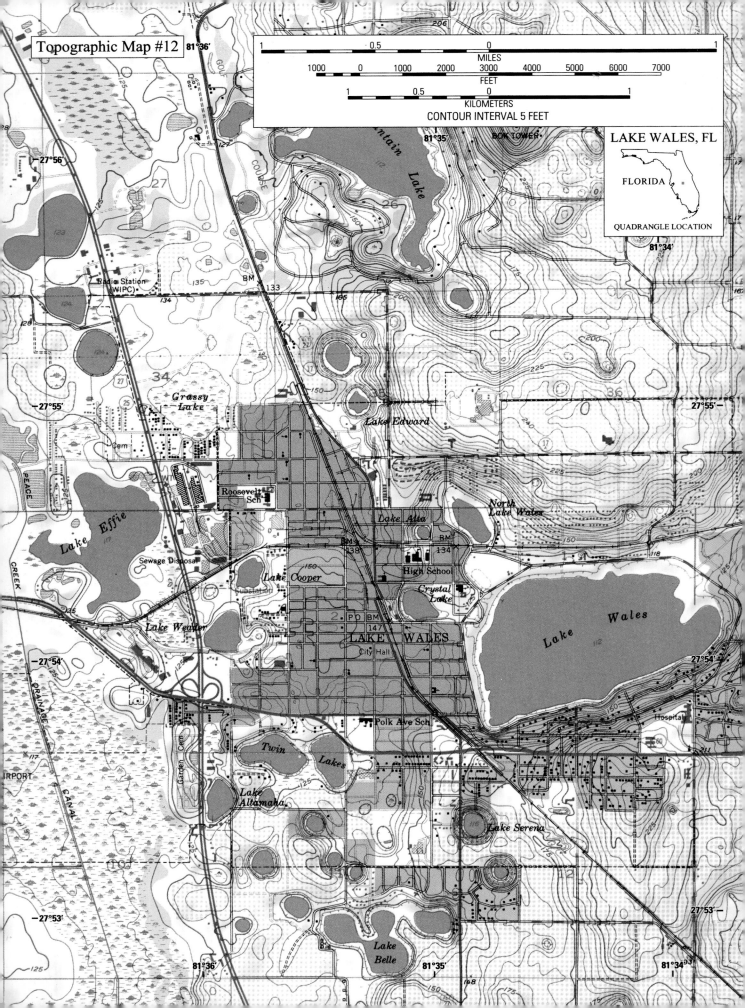